The Big Book of Drones

Drones are taking the world by storm. The technology and laws governing them change faster than we can keep up with. *The Big Book of Drones* covers everything from drone law to laws on privacy, discussing the history and evolution of drones to where we are today. If you are new to piloting, it also covers how to fly a drone including a pre-flight checklist.

For those who are interested in taking drones to the next level, we discuss how to build your own using a 3D printer as well as many challenging projects for your drone. For the truly advanced, the *Big Book of Drones* discusses how to hack a drone. This includes how to perform a replay attack, denial of service attack, and how to detect a drone and take it down.

Finally, the book also covers drone forensics. This is a new field of study, but one that is steadily growing and will be an essential area of inquiry as drones become more prevalent.

The Big Book of Drones

Ralph DeFrangesco
Stephanie DeFrangesco

CRC Press
Taylor & Francis Group
Boca Raton London New York

CRC Press is an imprint of the
Taylor & Francis Group, an informa business

First Edition published 2023
by CRC Press
6000 Broken Sound Parkway NW, Suite 300, Boca Raton, FL 33487-2742

and by CRC Press
4 Park Square, Milton Park, Abingdon, Oxon, OX14 4RN

CRC Press is an imprint of Taylor & Francis Group, LLC

ISBN: 978-1-032-06281-5 (hbk)
ISBN: 978-1-032-06282-2 (pbk)
ISBN: 978-1-003-20153-3 (ebk)

DOI: 10.1201/9781003201533

Typeset in Sabon
by KnowledgeWorks Global Ltd.

Contents

About this book xi

1 Introduction **1**

Construction 5
Cinematography 5
Agriculture 5
Entertainment 6
Save lives (drop life-saving ring to drowning victims) 6
Drone delivery 6
Finding people 6
Drone racing 7
Fun 7
Military 7
The drone market 8
Top five states that will benefit from drone manufacturing 9
DoD drone classes 9
Other classification systems 10
Summary 13

2 The history of drones **15**

Ancient times 16
1400s 16
1700s 16
1800s 17
WWI 18
WWII 19
The use of military drones today 22
Iran shoots down a US drone 22
Military drones used in other countries 22

Fixed wing 23
Remote controlled 23
Takeoff and landing 23
Which other countries own drones? 23
Countries that do not have drones 24
Iranian anti-drone capability 25
The history of the FAA 26
Summary 27

3 Laws governing drones 29

Airports 30
The White House 30
Sports arenas 31
Wildlife refuges 31
Forest fires 32
Military bases and Department of Energy (DOE) sites 32
Prisons and other correctional facilities 32
National parks 33
Schools 33
City parks 34
Large social gatherings 34
Local laws 34
Drone rules and regulations 35
Part 107 pilot's license 35
Register your drone 36
Drone accidents 37
Accidents/incidents 37
Reporting incidents 37
Government involvement 38
Drone insurance 38
State drone laws 39
Privacy 39
Privacy legal cases 40
International laws 40
Europe 40
Austria 41
France 41
Germany 41
Ireland 42
Italy 42
South and Central America 42

India 42
Southeast Asia 42
Africa 43
Oceania 43
Countries to avoid 43
Summary 43

4 Drone hardware/software 45

Consumer versus commercial drones 45
Consumer drone manufacturers 46
Fixed-wing drones 49
Underwater drones 50
Commercial drone manufacturers 50
Military drones 53
Drone costs 55
Consumer drone add-ons and accessories 56
Sensors 56
Troubleshooting 59
Testing with Linux 60
What if one or more motors are not spinning? 60
What if your camera does not work? 61
Analyzing the controller signal 61
Other troubleshooting 62
Summary 62

5 Flying a drone 65

Safety tips 66
Location 69
Spare parts 70
Commercial flying 72
Controller specifics 77
Drone incidents 77
UA visual perception 78
Military drone pilots 80

6 Hacking a drone 81

The Telnet protocol 82
Telnet to the Parrot AR 82
Transferring files to and from the Parrot AR 84
Getting the MAC address 85

Telnet to the Phantom 3 85
How to turn on Telnet 86
Transferring files to and from the Phantom 3 87
Other drones 87
Deauthenticating the Parrot AR 87
Replay attack 88
How to identify a drone 89
Approach 1 89
Approach 2 93
Taking down a drone 97
Protecting your drone 99
Protect the controller 99
Use a Virtual Private Network (VPN) 99
Use strong passwords 99
Use encryption 100
Buy a dedicated controller 100
Fly in a remote area 100
Be aware of your surroundings 100
Summary 101

7 Programing a drone 103

Code explanation 105
Using other languages 105
The code behind the project 107
Using Python to program a drone 108
Other ways to program a drone 110
Summary 112

8 Build your own drone 115

Drone racing 115
Videography 116
Do-It-Yourself (DIY) drone kits 116
3D printing a drone 117
Build your own underwater drone 121
A very cheap alternative 123
Summary 123

9 Do-It-Yourself (DIY) drone projects 125

What are some of accessories you can mount on your drone? 126
What can I use my drone for? 127

Drones to help the blind 127
Drones to watch your kids 128
Drones to watch your house 129
Personalized drone videos 130
Build a drone race course 130
Drone bug-out bag 132
Multiple drones 133
Drone versus drone 134
Drone fishing platform 135
A drone light show 136
Take your drone snowboarding 137
How to make money with a drone 137
Selling photos and videos 137
Wedding photography 137
Selling drones 138
Working with a realtor 138
Working as a drone pilot 138
Summary 139

10 Drone forensics **141**

Introduction to forensics 142
Let's analyze the data 148
Summary 149

11 More on drones **151**

Drone conferences 152
A closing note on drones 155
Register your drone 155
Weather 155
Flying at night 155
Weight they can lift 156
Battery length 156
Banned countries 156
Privacy 156
The future of drones 159

Definitions 163
Index 167

About this book

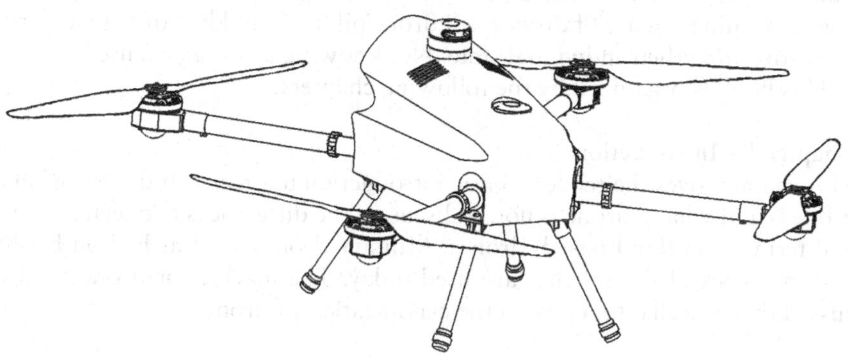

Image used from public domain: https://commons.wikimedia.org/wiki/File:EM120_drone.jpg
EM120 drone

Like so many other things in technology, necessity is the mother of invention. Three things happened that were the drivers for writing this book; first, this book started out as the two of us keeping notes. There were notes, more notes, and even more notes after that. Some notes were just reminders such as the preflight check material. Some were notes related to observations when flying. We transferred the many notes written down on pieces of loose paper to a notebook. Obviously, the notebook kept growing.

Second, many times when out flying our drones, people would stop and ask us questions. Most were basic questions about flying. Some were questions about drone-related laws, and some were hardware and software related. We started writing down the questions and how we responded. We would do a little research to make sure we were right in our responses.

Lastly, we noticed that people wanted to learn more about drones. Undoubtedly the internet is a quick go to solution, but the internet can only take you so far. We decided to start our own drone conference called USDroneCon. At USDroneCon, you can learn how to fly, or fly a drone you have never flown before. There are hands on labs and guest speakers that talk about their expertise and experiences they have with drones.

PREAMBLE

This book covers the history, laws, and uses of drones. It will explain how the military, commercial, and consumer sectors have adopted drones. Where applicable, the difference is notated and explained. Drones are a fast growing technology. The material presented in this book is the latest available; however, the technology is evolving so quickly that it could be outdated by the time this book is published. This is especially important when talking about laws governing the use of drones. All drone pilots should be familiar with the laws governing their individual state. Not knowing is not an excuse.

This book is organized by the following chapters.

Chapter 1 – Introduction

This chapter gives the reader a basic introduction to drones. It defines what a drone is and what a drone is not. It discusses the differences between air, sea, and terra-firma (land based) drones. Military drones are touched on briefly and the types of drones that are used today. The market for drones is discussed then the chapter ends on the classification of drones.

Chapter 2 – History of drones

This chapter discusses the history of drones from ancient times to today's drones and how the technology has evolved. This chapter touches on military drones in other countries, and some countries that do not allow the use of drones.

Chapter 3 – Laws governing drones

This chapter is important since it discusses the laws that govern the use of drones. Specifically, this chapter lists the places that drones are banned and why. The drone registration process is discussed in addition to drone insurance. The chapter ends with reported drone incidents that have occurred recently.

Chapter 4 – Drone hardware/software

Hardware and software are really the focus of drone technology. This chapter discusses how consumer drones differ from commercial and military drones. A few drone manufacturers and some of the drones they sell are introduced along with sensors and options available for drones.

Chapter 5 – Flying a drone
This chapter discusses the checklist that a drone pilot should use before flying a drone. The know where to fly list is discussed along with how to do stunts and how to set the auto land function.

Chapter 6 – Hacking a drone
Hacking a drone is different than hacking a computer. The processes and procedures may be similar, but the end result is very different. You will learn several ways that a drone can be hacked and what treasures the drone might hold!

Chapter 7 – Programming a drone
Not all drones can be programmed, but for the ones that can this opens up a whole new world for programmers. You will learn how to connect to your drone, send it off to do a mission and then return safely to you, all through programming.

Chapter 8 – Build your own drone
Most computer enthusiasts enjoy tinkering with hardware. People who fly drones, share the same enthusiasm. This chapter will show you how to build a drone from a kit and take you through 3D printing your own drone body.

Chapter 9 – Do-It-Yourself (DIY) drone projects
Have a drone and not sure what you can do with it? The uses for a drone are endless. This chapter shares several projects that drone pilots might want to tackle.

Chapter 10 – Drone forensics
Did the drone crash into the crowd on accident or on purpose? This chapter walks the reader through how to perform forensics on a drone. You will learn how to make a scientific guess at who owns that drone and how far it might have flown before it crashed. This chapter includes a forensic checklist to use in a forensic investigation.

Chapter 11 – More on drones
Do you need replacement parts for your drone? Want to add Lidar or a high-definition camera? This chapter includes links and information to companies that supply parts for drone enthusiasts.

Chapter 1

Introduction

Image used from public domain: https://unsplash.com/photos/L9wrEGJjRdo

Drone Image

Drones are not new by any means. They are, however, new to today's consumers. Drones date back thousands of years, where they were mostly used for military operations. They still play a major role today in military operations, but they also have found their way into the hands of an average consumer.

Drones are pervasive since they are cheap and readily available. Costs have plummeted to where a consumer can purchase a drone for as little as $20 or less. The capabilities will be limited at this price range, but it still functions

DOI: 10.1201/9781003201533-1

as a drone. If we look at the opposite end of the spectrum, you can easily drop $1200 or more on a consumer drone. Of course, a drone in this price range has multiple cameras, an active gimbal, GPS, and a dedicated controller, which we will explain more throughout this book.

Let's start our journey by asking, what is a drone? Drones are referred to by several names: Unmanned Aerial Systems (UAS), Unmanned Aerial Vehicles, Remotely Piloted Aircraft, and quadcopters. Whatever you call them, the basic concept is the same. The Federal Aviation Administration (FAA) defines a drone as a vehicle which is not piloted by a human from within the vehicle itself. This is a good start as a definition, but negates the fact that some specialized drones do use a wired tether by design. We will explain more about these drones later.

Whatever you call them, they all have one thing in common, there is no pilot in the vehicle, and the vehicles flight or maneuver is Remotely Controlled (RC), either directly by a pilot or preprogramed software. Drones that are made to go underwater are the exception. They can use wires for control and communications. This is necessary because a consumer underwater drone is not capable of receiving wireless signals through the water. Maybe they will get there someday.

Now that we have an idea as to what a drone is, well then, what is not considered a drone? Balloons are not considered drones, although history considers hot air balloons as the very first drones. Wind-up rubber band airplanes and kites are not drones either.

If we look at the above, none are piloted remotely with wireless technology. Although some people would argue that a kite is piloted remotely, it is controlled with a string which is not a remote-controlled technology, and you are limited in maneuverability by the string itself.

Now, let's take a look at the types of drones in the market. Drones that go underwater are in trend now. The limiting factor is the distance they can go and their cost as they tend to be quite expensive. As an example, the QYSEA FIFISH V6 Underwater ROV is a sea drone. This model sells for roughly $1300.00. It's capable of diving roughly 160 feet. It has a 4k camera, which can take video and 12MP photos. The QYSEA FIFISH V6 Underwater ROV has a top speed of 3 knots and the manufacturer claims it can stay submerged for up to 4.5 hours on a single charge.

Land drones or terra-firma drones have probably been around the longest. We have come to know them as RC cars. In general, RC cars can reach a top speed of 46 km/h. They can last roughly 40 minutes on one battery charge or longer if there are extended battery options. The RC car below is charged with a Universal Serial Bus (USB) cable, but other cars have batteries that are removable which means you can be using the car while charging another set of batteries.

Image used with permission from Stacy and Andrew Yeager
Yeager, S. & Yeager, A. (2021). Remotely Controlled (RC). Bloomsburg, PA.

When people typically think about a drone, they think about a drone that flies in the air. The picture below shows a typical quadcopter. The model shown has a camera mounted to a gimbal. The quadcopter shown below has a 20-minute battery life, a separate dedicated controller, and is capable of aerial stunts.

Image used with permission from Ralph M. DeFrangesco
DeFrangesco, R. (2021). Quad-copter Williamstown, NJ

A fixed-wing drone is the newest family of consumer drones that fly in the air. Typically, you have to throw them in the air in order to get them airborne, and then take over flying them. However, some do have propellers to assist with takeoff. The fixed-wing drone below would need to be thrown into the air for takeoff. It does have a landing gear, but it's very impractical when landing. You would need a perfect landing surface along with a perfect landing in order for it to land on the wheels. When flying this type of drone, landing is more of a controlled crash.

Image used with permission from Ralph M. DeFrangesco
DeFrangesco, R. (2021). Fixed Wing Drone Williamstown, NJ

We now have a definition for a drone and have examples of drones. All of these have been in regard to consumer drones. Let's see how the military defines a drone. According to the military, a drone is a land, sea, or an air vehicle that is remotely piloted or automatically controlled. In this book, we will only touch on military drones that fly.

The military has many different types of drones to meet their mission requirements. The MQ-1B Predator is operated by the US Airforce. This drone is capable of flying on a long range, long endurance mission. Predator missions might include reconnaissance, surveillance, close air support, search and rescue, precision strike, convoy overwatch, and terminal air guidance. The MQ-1B is manufactured by General Atomics Aeronautical Systems and costs roughly $20 million dollars depending on its configuration.

The MQ-9 Reaper is operated by the US Airforce. The Reaper is primarily an attack platform and can be used as an intelligence collection tool.

The Reaper utilizes a laser range finder and a synthetic aperture radar. The platform is capable of carrying four laser-guided Hellfire missiles which can be deployed in anti-armor and anti-personnel engagement missions.

The RQ-4 Global Hawk is operated by the US Airforce. The Global Hawk is a high-altitude and high-endurance reconnaissance platform. Northrup Grumman is the prime manufacturer. By now, we should now have a pretty good idea as to what a drone is and some examples of drones. However, what can drones be used for?

CONSTRUCTION

The construction industry is one of the largest users of drones. They use commercial drones primarily for inspection purposes. A drone can get into places a human cannot, or at least to places that are not very easy to go, without any risk to human life. As an example, a bridge inspection company can send a drone to look under a bridge, checking the integrity of its beams and girders. Drones have cameras, so they can inspect and document at the same time without putting a human at risk. If something is found, then a human can be used to verify the findings of the drone.

CINEMATOGRAPHY

The use of drones in cinematography has brought about new perspectives for the movie-making industry. Action shots that were either not possible or costly to make have been easily made with drone technology at a fraction of the cost. Drones have replaced humans in making dangerous video shots lessening the threat to human life.

The first major movie to use a drone for film scenes was Skyfall in 2012, starring Daniel Craig, Javier Bardem, and Naomie Harris. The director decided to use a drone rather than a traditional helicopter to film the aerial footage.

Other films that used a drone were The Expendables 3 (2014), The Wolf of Wall Street (2013), and Chappie (2015).

AGRICULTURE

Although not the largest users of drones, the agricultural industry uses commercial drones in a big way. Drones have been used to help farmers for many years. Sensors mounted on drones have been used to track crop diseases, create irrigation maps, and track the crop output. This advanced technology has helped farmers to increase crop yield and produce crops resistant to bugs and diseases.

ENTERTAINMENT

Drones purchased by consumers are bought for various reasons. Parents buy drones for their children to be used for entertainment. Adults purchase drones for novelty purposes. The witted homeowner will occasionally use a drone to check that their gutters are not clogged or for worn or damaged shingles. For the most part, consumers have no "need" for a drone except to have fun with.

SAVE LIVES *(DROP LIFE-SAVING RING TO DROWNING VICTIMS)*

Research work has been done with drones to see if they can detect if a person is in distress and might need help from drowning. Drone technology can determine that a person is thrashing around in the water and waving their hands. This might be an indicator that they need help. A life preserver could be dropped from the drone until help arrives.

DRONE DELIVERY

In 2013, Jeff Bezos, the CEO of Amazon, announced that they would start utilizing drone technology to assist in the delivery of small payload packages. In 2016, Amazon made good on its promise successfully delivering a package to an Amazon Prime customer in the United Kingdom.

Amazon started Amazon Prime Air delivery service in order to fulfill this business model. The idea was to be able to deliver packages weighing less than 5 pounds in less than 30 minutes. The delivery address cannot be more than 10 miles from any Amazon fulfillment center. In 2020, the Amazon drone fleet did receive FAA clearance to begin its delivery service. As of December 2020, Amazon has yet to deliver any packages on a commercial basis using this service except the initial package delivered in 2016.

Another company using drones for delivery is Dalsey, Hillbloom and Lynn (DHL). DHL China is delivering small packages weighing less than 12 pounds and within a 5-mile radius, in less than 8 minutes.

In April 2019, Google received permission from the FAA to start testing drone deliveries in the United States. In October 2019, Google delivered its first package via drone for Pharmacy retailer Walgreens. We will see who wins the drone delivery war: Google or Amazon?

FINDING PEOPLE

In July 2019, a family was joining in a community celebration in Westerville, Ohio. While enjoying the activities, a father approached the Fire Chief explaining to him that his nine-year-old daughter was missing. The Westerville fire company and police had just spent $34,000 on a drone as part of their new drone program. The police immediately put the drone in flight to find the girl. The drone spotted her within 10 minutes. She had walked back to the family car where she was sitting on the ground. The drone spotted her quickly because of the detailed description the family gave of her and the high-definition cameras on the drone.

In November 2019, the police in Sun Prairie, Wisconsin, used a drone equipped with thermal imaging to find a missing man. The man was thought to have had a medical emergency while driving his car. The man drove his car into a field and wondered out. A police officer found his car running by the side of the road without anyone around it. They called in a drone to try and find him. The drone picked up a thermal signature of the man within 2–4 minutes.

DRONE RACING

Drone racing is a fairly new sport that started in Australia in 2013. The pilots use First Person View (FPV) to observe what the drone sees. Drones have a camera that transmits images from the drone to goggles that the pilot is wearing. Any drone can be fitted with FPV, but many pilots utilize high-quality, after-market gear that can cost thousands of dollars.

The Drone Racing League (DRL) was founded in the United States in 2015. The DRL operates internationally airing professional racing on National Broadcasting Company (NBC) Sports, Sky Sports, ProSieben, Groupe AB, and Fox Sports Asia. There are many racing leagues that host drone racing:

- MultiGP
- DR1 Racing
- RotorMatch League

Each organization sets the standards for their racetracks. This could include the size (length), obstacles, indoor versus outdoor, and track location. A lot of racing is done in stadiums because of the possibility of bad weather.

FUN

People just like flying drones and the most amount of fun you can have is using drones to do tricks. It tests your skill and agility along with the drone's capabilities.

MILITARY

The US military has used drone technology for over a hundred years. Drones were used for observation in the beginning. They have been used to drop bombs and spy on other armies. Today, they are used as sophisticated missile platforms and for reconnaissance purposes.

THE DRONE MARKET

The following graphic shows the worldwide sales growth projection for drones through 2030. In a whitepaper from Levitate Capital (2020), the largest market segment in 2020 was the defense market. However, Levitate Capital expects that the commercial market will surpass the defense market by 2025.

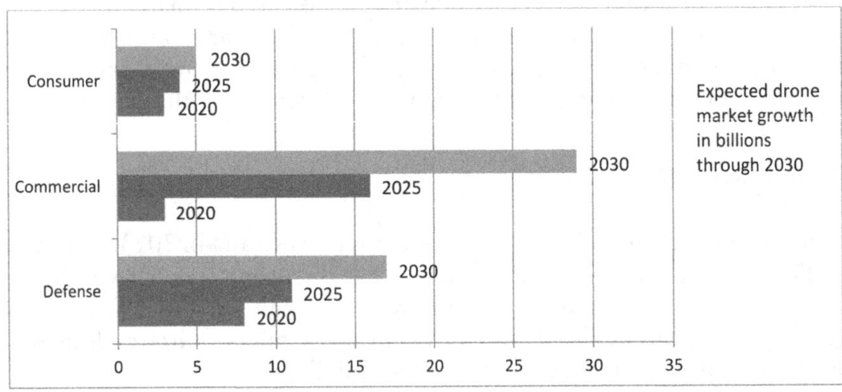

The following two graphics show how the commercial market is broken down in sales from 2020 to 2025. As we can see from the chart, the largest growth is expected in the construction industry followed by inspection and agriculture.

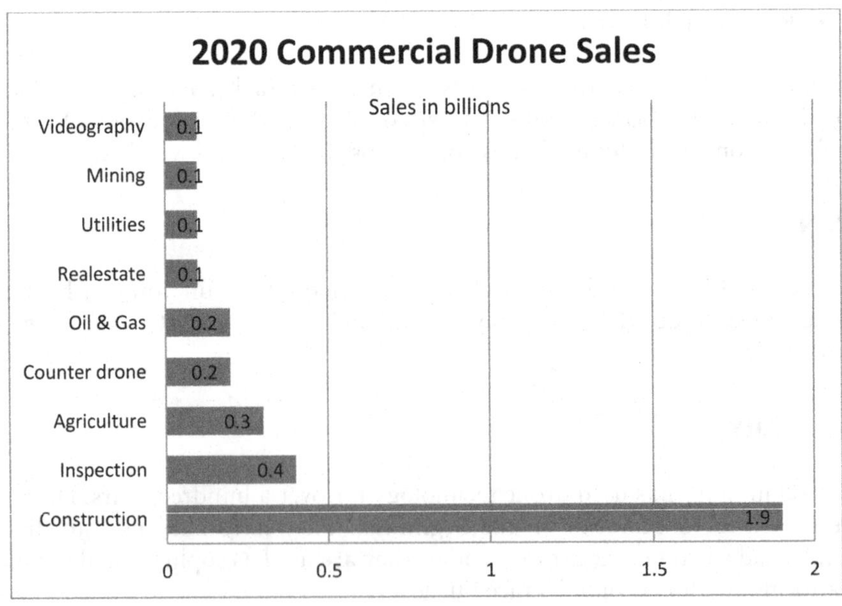

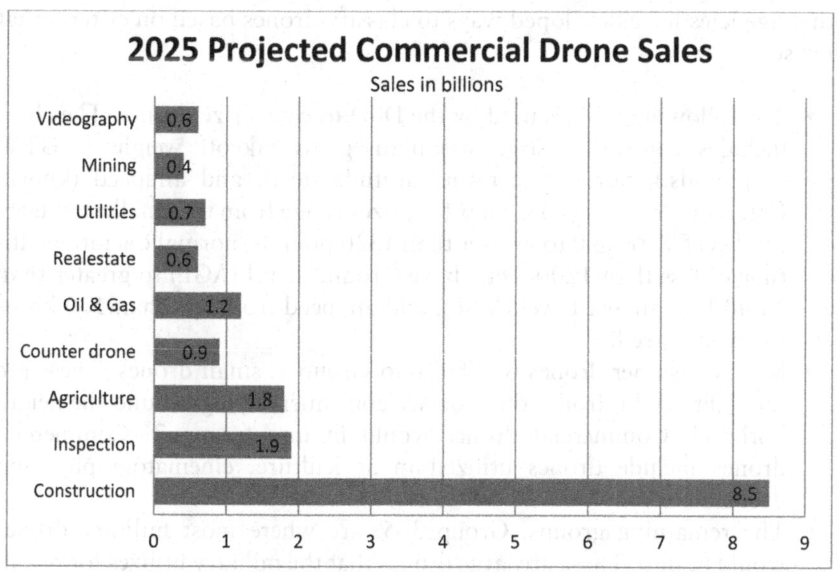

TOP FIVE STATES THAT WILL BENEFIT FROM DRONE MANUFACTURING

As you would expect, with this much growth on the horizon, someone will benefit from it. The chart below shows which states will benefit the most from drone production.

State	*Number of jobs*	*Total economic impact*
California	2108	400M
Washington	1157	218M
Texas	958	181M
Florida	557	105M
Arizona	494	93M

As we can see, California is one of the top states. This is not surprising given they are such a high-tech state. Washington, Texas, Florida, and Arizona complete the list.

DoD drone classes

Not all drones are created equal. Certainly, consumer drones have a different function than commercial and military drones. They differ in size, shape, capability, and functionality. The Department of Defense (DoD) and other agencies have developed ways to classify drones based on criteria that they set.

- The following table is used by the DoD to categorize drones. The chart includes categories, size, maximum gross takeoff weight (MGTW in pounds.), normal operating altitude (feet), and airspeed (knots). Categories range from Group 1–5, sizes range from the smallest to largest, MGTW from 0 to greater than 1320 pounds, normal operating altitude of less than 1200 feet Above Ground Level (AGL) to greater than 18,000 Mean Sea Level (MSL), and airspeed from less than 100 knots to any airspeed.
- Most consumer drones will fall into Group 1, small drones. These are the lightweight drones that you see consumers flying around the neighborhood. Commercial drones would fit into Group 2. Commercial drones include drones utilized in agriculture, cinematography, and drone delivery.
- The remaining groups, Group 3–5, are where most military drones would fit into. These are large drones that the military utilizes for reconnaissance, troop protection, scouting, and missile deployment. These drones can cost millions of dollars not including their payloads. They are RC with the pilot sitting thousands of miles away.

Category	*Size*	*Maximum gross takeoff weight (pounds)*	*Normal operating altitude (feet)*	*Airspeed (knots)*
Group 1	Small	0–20	<1200 AGL	<100
Group 2	Medium	21–55	<3500	<250
Group 3	Large	<1320	<18,000 MSL	<250
Group 4	Larger	>1320	<18,000 MSL	Any airspeed
Group 5	Largest	>1320	> 18,000	Any airspeed

Note: If the UAS has even one characteristic of the next level, it is classified in that level.
AGL: Above Ground Level.
MSL: Mean Seal Level.

Other classification systems

There are a number of other drone classification systems. The North Atlantic Treaty Organization (NATO) uses a similar classification system to the DoD system. Class and weight in kilograms, category and weight in kilograms, normal operating altitude in feet, normal mission radius in kilometers, and an example platform are all similar.

- Drone classes range from Class I to Class III
- Weight is from 150 kilograms to 600 kilograms

- Normal deployment can be tactical, strategic, or operational
- Normal operating altitude ranges from less than or equal to 5000 feet–65,000 feet
- Normal mission radius ranges from 5 kilometers to unlimited
- Some examples shown in the chart are the Luna, Skylark, Hermes 450, Global Hawk, and the Predator

Class and weight (kilograms)	*Category and weight (kilograms)*	*Normal employment*	*Normal operating altitude (feet)*	*Normal mission radius (kilometers)*	*Example platform*
Class I <150	Small >20	Tactical Unit	<5000 AGL	50 (LOS)	Lina, Hermes 90
	Mini	Tactical Unit	<3000 AGL	25 (LOS)	Scan Eagle, Skylark, Raven
	Micro	Tactical Patrol/ Section (single operator)	<200	5 (LOS)	Black Widow
Class II 150–600	Tactical	Tactical Formation	<10,000	200 (LOS)	Sperwer, Iview 250, Hermes 450
Class III >600	Strike/ Combat	Strategic/ National	<65,000	Unlimited (BLOS)	
	HALE	Strategic/ National	<65,000	Unlimited (BLOS)	Global Hawk
	MALE	Operational/ Theater	<45,000	Unlimited (BLOS)	Predator A & B, Heron, Hermes 900

Note:
LOS: Line of Sight
MSL: Mean Sea Level
BLOS: Beyond Line of Sight
AGL: Above Ground Level
HALE: High Altitude Long Endurance
MALE: Medium Altitude Long Endurance

If we compare the NATO table against the DoD table, the main difference is drone weight by class. As an example, the smallest drone in the Group 1 category of drones from the DoD table is 0–20 pounds. However, if we look at the NATO table, the smallest drones are in Class 1 weighting less than 150 kilograms, which is roughly 300 pounds. They do break this category down further into three additional weight classes.

Operation			UAS				
Subcategory	*Area of operation*	*Remote pilot competency*	*Class*	*MTOM/ Joule (J)*	*Main technical requirements*	*Electronic ID/ GEO awareness*	*UAS operator registration*
A1 Fly over people	You can fly over uninvolved people (not over crowds)	Read consumer info	C0	<250 grams	Consumer info, Toy Directive or <19 m/s, no sharp edges, selectable height limit	No	No
		Consumer info, online training, online test	C1	<80 Joule or 900 grams			
A2 Fly close to people	You can fly at a safe distance from uninvolved people	Consumer Info, Online training, online test, theoretical test in a center recognized by the aviation authority	C2	<4 kilograms	Consumer info, mechanical strength, no sharp edges, lost-link management, selectable height limit, low-speed mode	Yes, unique SN for identification	Yes
A3 Fly far from people	Fly in an area where it is reasonably expected that no uninvolved people will be endangered and keep a safety distance from urban areas	Consumer info, online training, online test	C3, C4, privately built	<25 kilograms	Consumer info, lost-link management, selectable height limit	If required by zone of operation	Yes

The European Union Aviation Safety Agency (EASA) differs drastically from the DoD and NATO classification systems. EASA only classifies consumer drones. The EASA standard covers subcategories, area of operation, remote pilot competency, class, maximum takeoff mass, main technical requirements, electronic ID/Geo awareness, and UAS operator registration.

Interestingly enough, in subcategory A1, the EASA framework allows a drone to fly over people. The drone can only fly over uninvolved people, or people that are not part of the pilot's team. However, in no circumstances can it be flown over a crowd. The other important EASA requirement is that drones that fall into class C1, C2, and C3 must broadcast an E-Identification signal. This signal broadcasts an ID and the takeoff location of the drone.

SUMMARY

This chapter was a basic introduction to drones. We covered what the definition of a drone is. We discussed the different types of drone that are there: air, Terra-firma, and underwater. We then discussed some of the key uses for drones; these include, but are not limited to cinematography, agriculture, package delivery, finding people, personal entertainment, and professional drone racing.

With the sales of drones increasing yearly, we discussed how much is expected to be spent over the next few years, what industries are spending the money, and what states will benefit from the manufacture of drones.

Finally, we discussed drone classification systems. There are many classification systems. We discussed three of the major ones used worldwide: DoD, NATO, and EASA. Now that we have a good understanding of what drones are and how they can be used, Chapter 2 will discuss the history of drones. What were the first drones? What were they used for? What do drones look like today and what does the military use them for? All of these questions and more relating to the history of drones will be covered in Chapter 2.

Chapter 2

The history of drones

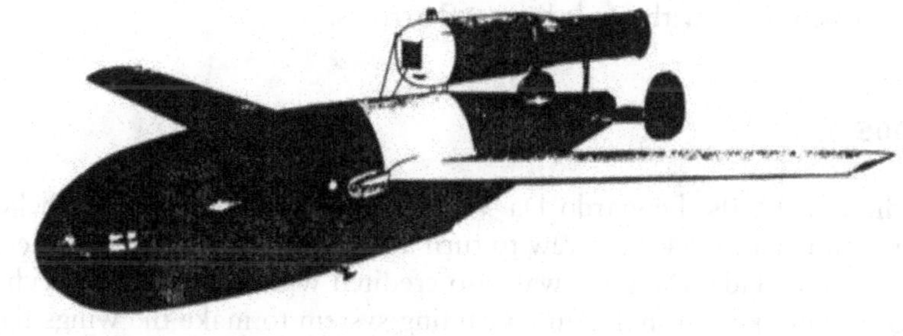

Image used from the public domain: https://commons.wikimedia.org/wiki/File:Globe_KD2G_Firefly_SAC.jpg
Globe KD2G Firefly SAC

You may be wondering why drones are called drones? The use of the word has an interesting connotation. A male bee has one job; to mate with a queen bee. This is obviously a luxury job and is considered idle work as compared to a worker bee. So this type of bee is known as an idler or drone, hovering around waiting to mate with a queen bee.

As we mentioned in Chapter 1, balloons are not really drones, but history does consider their use as "drone like". They are not really piloted remotely but can deliver a payload if needed. We have seen this throughout ancient history where people have used balloons for delivering fire bombs to try and destroy distant cities. This chapter looks at the history of drones from ancient times through modern times.

DOI: 10.1201/9781003201533-2

ANCIENT TIMES

The use of unmanned aircraft has been traced back to ancient Greece. Supposedly, Archytas, an inventor from the city of Tarantas, built a mechanical bird that was steam powered. Folk lore states he flew his mechanical bird 200 meters, but this cannot be confirmed through historical records or facts.

Around the same time, the Chinese were experimenting with crude aircraft technologies. They developed hot air balloons and kites they used for military purposes. They managed to figure out how to "bomb" their enemy from the air. Although not technically like drones, but they had the same effect on the enemy. Of course, one weakness with hot air balloons is that they can be shot down easily, even with crude bow and arrows.

1400S

In the late 1400s, Leonardo Da Vinci put his ideas for the first helicopter down on paper. It used a screw to turn a shaft, which in turn provided rotation for the blade. Da Vinci was also credited with developing a mechanical bird that used crankshafts and a cabling system to make the wings flap. Da Vinci's ideas died on paper as he never built either design. One could only imagine that if he built either of these, how further along the technology might be.

1700S

The 1700s were a very interesting time in drone history. Commercial balloons came into existence. We begin to see aviation pioneers like the Montgolfier brothers, Carl Friedrich Meerwein, and Jean-Pierre Blanchard experimenting with balloon technology and various flying machines. Meerwein, as an example, was a university graduate with a background in mathematics, physics, and engineering. He graduated from the University of Strasbourg with a degree in civil architecture. Meerwein loved the idea of flying. In 1781, he designed and supposedly flew an ornithopter. This is a bird-like device that gets its lift by flapping its wings under human power. However, attempts by Meerwein in 1784 and 1785 failed, which put in doubt regarding his first alleged successful attempt.

Taken from the public domain: https://commons.wikimedia.org/wiki/File:Ornithopter_(PSF).png
An Ornithopter

The above is an example of the ornithopter that Meerwein developed in 1781. The idea is that a person would strap themselves into the device and begin flapping their arms causing the device to flap its wings that provides lift.

1800S

Nikola Tesla was a pioneer in remote control technology. He demonstrated that he could remotely control a boat with electricity in 1898. Little did Tesla know that the technology he just unleashed would be used by the military to control machines of war.

Roughly in 1849, the Austrian military used unmanned balloons filled with explosives to bomb Venice, Italy. The plan had some success until some of the balloons blew back into Austria. What this showed is that it was possible to bomb a country miles away and cause fear in people.

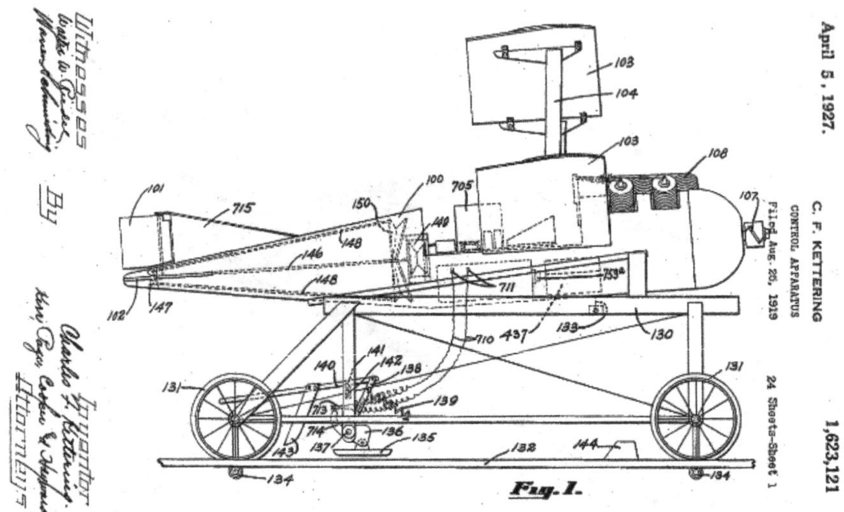

Kettering, C. F. (1919). Patent No. US1623121A.
Kettering Bug Blueprint

WWI

Both the United States and the UK developed drones during WWI. The United Kingdom developed the Aerial Target and the United States developed the Kettering Bug. Both were considered radio controlled aircrafts.

The above picture is the control apparatus for the Kettering Bug. The "Bug" was also known as an aerial torpedo. As we can see in the picture, it was a rather fragile looking aircraft. It had no landing gear as we are familiar with today. It used skid bars as a type of crude landing gear. It could strike targets up to 75 miles away and reach speeds of 50 mph. It was developed by Dayton-Wright Airplane Company. Orville Wright was a consultant on the project. The Bug was essentially made of paper and wood and cost only $400.00 at the time to produce.

The Bug was modified several times. The below picture is one of its many iterations. This one included steel wheels as a landing gear along with a bi-wing. This version was a little larger and heavier than its predecessors but was capable of carrying a larger payload. It is shown below on a set of metal rails to help it to takeoff.

Taken from the public domain: https://en.wikipedia.org/wiki/Kettering_Bug
An iteration of the "Bug"

WWII

The actor Reginald Denny (see picture below), also known as Reginald Leigh Dugmore, was an aviator, inventor, and Unmanned Aerial Vehicle (UAV) pioneer from the 1920s through the 1950s. In 1934, Reginald Denny started selling radio controlled planes under the brand Reginald Denny Hobby Shops. Denny was able to show the US military that they could be used as target drones. During WWII, the Unite States converted a used B-17 bomber into a radio controlled weapon. The bomber was packed with explosives and then using radio control technology, the bomber was guided to its target.

Taken from the public domain:
https://en.wikipedia.org/wiki/Reginald_Denny_(actor)#/media/File:Reginald_Denny_in_Stars_of_the_Photoplay,_1924.jpg
Reginald Denny

In the 1950s, the United States continued with developing drone technology. The F6F-5K Hellcat drone carried a 100-pound bomb. A radio repeater was used to guide the bomb to its final destination. The F6F-5K is pictured below in a non-drone configuration.

Taken from the public domain: https://commons.wikimedia.org/wiki/File:Grumman_F6F-3_Hellcat_of_VF-1_in_flight_over_California_(USA),_in_1943_(80-G-K-605).jpg
The F6F-5K Hellcat

On August 16, 1956, the Hellcat drone was involved in a rather ugly incident. Shortly after the drone took off from the Naval Air Station Point Mugu in California, the operator lost control of the drone. Luckily the drone did not have any ordinance onboard. The Navy, who was running the exercise at the time, scrambled two fighter jets from Oxnard Air Force base a few miles away. The jets engaged the runaway drone and fired several times at it with on-board rockets. Unfortunately, they missed. The jets then unloaded their entire ordinance at the drone, while some rockets did hit the drone, they failed to detonate. The drone finally ran out of fuel and crashed in the desert near Palmdale Regional Airport. The drone destroyed several power lines and caused numerous fires.

During the 1960s through mid-1970s, the United States used the Lightening Bug. This was also known as the Ryan Model 147, developed by Ryan Aeronautical. It was a small, jet powered, low-altitude drone that was primarily used for reconnaissance missions. The data collected from this drone was used for battle planning and combing damage assessment.

Taken from the public domain: https://en.wikipedia.org/wiki/Ryan_Model_147

Lightening Bug

As we can see from the photograph above, the Lightening Bug was deployed without landing gear. The drone was launched from the ground or from an aircraft. At the end of its mission, the drone fell back to earth using a parachute deployed by the operator.

THE USE OF MILITARY DRONES TODAY

Although they do a somewhat similar job, the drones of today differ significantly from the drones of 20 years ago. The military drones today are faster, larger, can carry more weapons, fly longer and farther, and they are remote controlled from across the world. Today's drone pilots sit in sophisticated command posts with another pilot and ordinance specialist. Most of the time, these command posts are not even located where the drones takeoff and land.

As part of their reconnaissance mission, drones carry sophisticated video recording and photographic systems. The equipment today is so sensitive it can easily take photographs that can be used to identify a person or even a license plate.

IRAN SHOOTS DOWN A US DRONE

In June 2019, the Iranian Islamic Revolutionary Guard Corps shot down an RQ-4A Global Hawk BAMS-D surveillance drone owned by the United States. The Iranian government claimed that the drone invaded Iranian airspace near Kuhmobarak in the southern province of Hormozgan. The United States claimed that it was in the international airspace over the Strait of Hormuz and had the proper authority to operate in that area.

The Iranian government claimed that two other drones have crashed on Iranian soil, both in 2005. In 2011, Iran claims to have brought down a Central Intelligence Agency (CIA) owned drone, which is believed to have jump started Iran's drone program.

MILITARY DRONES USED IN OTHER COUNTRIES

The United States pioneered drone technology, but by no means is it the only country to have drones. The United States frequently sells its technology to its allies. Drone technology is a hot item and is no different in this regards. However, they frequently do not sell all of the bells and whistles along with the drone. After all, you can't give it all away!

Whether they are US built or foreign built, most military drones are designed to do two things; reconnaissance and surveillance, and/or act as a weapons delivery platform. All military drones have a few things in common.

The vast majority of military drones are turboprop powered. This is for a few reasons:

- The lag time in controlling them remotely.
- The video would be hard to see at high speeds.
- Dropping ordinance at high speeds.

Fixed wing

- The vast majority are fixed wing.

    - Some research has been done on multi-rotor.
        - MQ-8B Fire Scout
        - Schiebel S-100

Remote controlled

Today, drones are controlled from across the world with control signals bouncing off of the satellites. There is no need for a pilot to be located where a drone either takes off or lands. Controlling a military drone looks very much like a video game display.

TAKEOFF AND LANDING

Manually operated drones need roughly a 5000 feet runway to takeoff and land. Automated drones can takeoff and land in as little as 3000 feet Crews maneuver the drones onto the runway and the pilots take over control from there. When a drone finishes its mission, flight crews recover the drone and its payload if it's a drone performing reconnaissance.

WHICH OTHER COUNTRIES OWN DRONES?

The following is a short list of countries that own combat drones. Countries may operate more than one drone type, so only primary drones are shown for each country.

- Brazil – Elbit Hermes 450

    - The Elbit Hermes 450 is manufactured by Israel. It's an older platform that is capable of reconnaissance, surveillance, and communications relay. The platform costs roughly $2M.

- China – Chengdu Wing Loong I, CH-3, CH4
  - The Wing Loong lines of drones are manufactured by the Chengdu Aircraft Industry Group in China. The primary operation of the Wing Loon is reconnaissance; however, the aircraft can be fitted with an air-to-surface missile. The Wing Loong is capable of carrying a 2200 pound air-to-surface weapon with a maximum service ceiling of 26,000 feet and 20 hours of flight time.
- India – IAI Harop (IAI Harpy 2)
  - The IAI lines of drones are manufactured by Israel. This specific drone is capable of homing in on radio emissions. When the IAI Harop detects an emissions signature and locks in on it, then heads straight for it and explodes on contact. The Harop can either operate autonomously or with human intervention. The Harop can carry a 23 kilograms warhead, fly for 6 hours, and cover 1000 kilometers on one fueling.
- Iran – Shahed 129
  - The Shahed 129 is manufactured by Shahed Aviation Industries. The drone is primarily a reconnaissance platform. The Iranians tried to mount a missile on the drone, but it failed as an attack platform. The Shahed can carry a 400 kilograms payload. It's capable of flying for 24 hours with a service ceiling of 24,000 feet.
- Russia – Sukhol S-70 Okhotnik
  - The S-70 Okhotnik is manufactured in Russia jointly by Sukhol and MIG. The S-70 is capable of performing a reconnaissance and attack mission. The S-70 can carry two internal weapons up to 2000 kilograms each. It has a speed of 620 miles per hour with a range of 6000 kilometers.

COUNTRIES THAT DO NOT HAVE DRONES

Believe it or not, there are several countries that have banned the use of drones. You may ask yourself why? These countries just do not know how to deal with the laws and controversies of flying drones. Developing polices to address privacy, injuries, legal issues, and insuring a drone, all take time. Many countries just don't want to address them right now, so they banned the use of drones. The following are countries that have banned their use:

- Algeria
- Barbados
- Brunei
- Cote d'Ivoire
- Cuba
- Iran (military intelligence shows this to be false)
- Iraq
- Kenya
- Kuwait
- Kyrgyzstan
- Madagascar
- Morocco
- Nicaragua
- Saudi Arabia
- Senegal
- Sri Lanka
- Syria
- Uzbekistan

The one notable country here is Iran. Even though they outlawed the use of drones, the government does have a substantial reconnaissance and combat drone program. The Iranian arsenal has been under development since the Iraq-Iran war in the early 1980s. Early drones were called Mohajer. They were crude frames with stationary cameras. Today, the Mohajer-6 can be equipped with a laser guided missile, precision striking bombs and sophisticated optical equipment.

Other Iranian drones include the Ababil, Karrar, Shahed-129, Saeqeh, Yasir, and the Kaman-12, their newest in the drone arsenal.

Iranian anti-drone capability

In 2011, the Iranians claim to have taken down a Lockheed Martin RQ-170 Sentinel drone. The Iranian government claims to have forced the drone down by its cyberwarfare unit. The US government claims the drone was shot down.

Also of note was that in 2019, Iran shot down the US drone RQ-4A Global Hawk surveillance drone. The Iranians claim the drone invaded their airspace and had the right to shoot it down. The matter is still disputed today.

The United States had a ban on the international sales of Unmanned Aerial Systems until 2019. The Trump administration changed the policy to allow the sale of military and commercial drones to just about any country that wants one with minimal restrictions. Armed drone platforms are subject to assessments under the Conventional Arms Transfer Policy and the DoD. There are no restrictions on the sale of consumer drones.

THE HISTORY OF THE FAA

It would be remiss of us if we didn't discuss the history of the Federal Aviation Administration (FAA) since they are the federal agency tasked with determining the laws regulating drones. The origin of the FAA dates back to 1938 when President Franklin Roosevelt signed the Civil Aeronautics Act. This legislation created a three member board to oversee accident investigations. In 1940, President Roosevelt split the Civil Aeronautics Authority (CAA) into two agencies: the Civil Aeronautics Administration and the Civil Aeronautics Board (CAB). The CAA was responsible for air traffic control, pilot, and aircraft certifications. The CAB was responsible for investigating accidents and general air safety.

In 1958, President Eisenhower signed the Federal Aviation Act, which created the Federal Aviation Agency. All duties performed by the CAA were transferred to the FAA. Retired Air Force general Elwood "Pete" Quesada became the first FAA administrator.

In 1963, President Johnson was concerned about the state of aviation, created the Department of Transportation (DOT) and moved the FAA under that authority. In doing so, its name was changed to the Federal Aviation Administration (again FAA). The accident investigation duties were transferred to the National Transportation Safety Board.

In 1970, the FAA created the Air Traffic Organization (ATO). The Air Traffic Control Systems Command Center (ATCSCC) is the operational arm of the ATO and is responsible for real-time command and control of the nation's airspace. According to the ATO's website, their functions include:

- Aligning data-driven changes in the operation
- Modernizing training to educate and prepare the ATO's technical workforce by focusing on timely delivery, operational priorities, and competencies
- Monitoring mitigations and changes to the nation's air space
- Informing stakeholders of successful strategies to meet safety goals

The ATO also provides a number of safety functions, which include:

- Event investigation
- Data analysis
- Corrective actions
- Training for over 14,000 air traffic controllers and over 6000 airway transportation system specialists
- Policy development, performance measurement, and promotion of a positive safety culture

The ATO realizes the need to address our nation's air space congestion and expansion. The ATO is developing the Next Generation Air Transportation System (NextGen). NextGen is an attempt to take US aviation into the 21st century and beyond. The system will be redesigned from the ground up addressing air gridlock, improvements in communications, satellite navigation, and shifts some of the decision making from the ground to the cockpit.

SUMMARY

This chapter really describes the history of drones. We covered the earliest drones all the way through today's military drones. We discussed which countries are known to use drones and which countries claim not to have any drones. There is disagreement about some of the countries on the list. Finally, we covered a thorough history of the FAA, since they play a critical part in the regulation of drones in the United States.

As a drone pilot, Chapter 3 is probably one of the most important. The chapter discusses drone laws in the United States, as well as in other countries. Specifically, it discusses places in the United States where you are not allowed to fly. There are many places where you are not permitted to fly your drone including airports, military bases, and the Whitehouse just to name a few. Local laws are discussed along with roles and regulations. Notable drone incidents are highlighted where drones have caused issues. One thing you will want to consider is drone insurance. We discuss what it is and how you can get it. Privacy is a big factor regarding drones today along with notable incidents involving drones and privacy are presented. Finally, international laws along with countries that do not welcome drones in their countries are listed.

Chapter 3

Laws governing drones

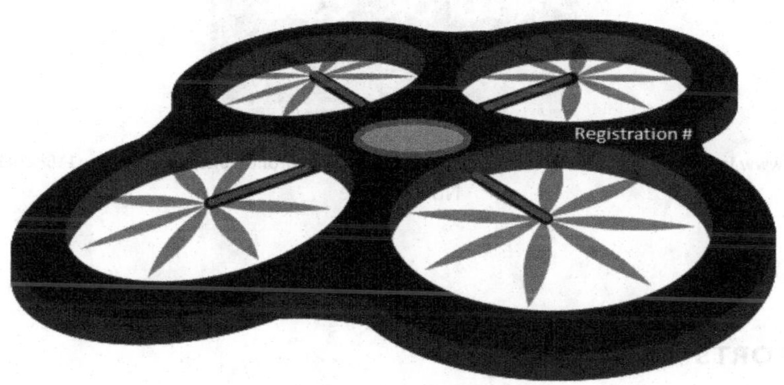

Image Modified from the public domain: https://commons.wikimedia.org/wiki/File:Quadcopter_Drone.png
Quadcopter Drone

As we have discussed in the previous chapter, the Federal Aviation Administration (FAA) is the governing body in the United States when it comes to regulations on drones. We will see in later chapters how states have enacted additional laws to combat privacy and using drones for things like hunting.

When drones first appeared on the market, there were a few restrictions. However, as the market increased and more and more consumers purchased drones, the FAA imposed flight restrictions. Today, drones cannot be flown just anywhere. The following is a list of places where drones are prohibited or at least with restrictions imposed. As always, it's best to check the FAA site for updated laws and your state for any additional restrictions.

Check the B4UFLY application to get more information on where you should and should not fly your drone. This site is updated regularly and should be your definitive guide on restrictions on where not to fly your drone. The application is available for free at the App Store for IOS and the Google Play Store for Android.

DOI: 10.1201/9781003201533-3

Taken from the public domain:
https://www.faa.gov/uas/resources/community_engagement/no_drone_zone/media/NDZ__1168x1415.png
No Fly Zone

AIRPORTS

It is not legal to fly within 5 miles of an airport. Any mishap could cause a plane to crash, and no responsible drone pilot would want to be a part of that. Legally, in the United States, drones cannot be flown within five miles of an airport. It is best to check which airports are near you before flying any drones.

At Gatwick Airport near London, between December 19 and 21 of 2018, hundreds of flights were canceled due to a drone interfering with flight operations. The authorities reported that approximately 140,000 passengers were inconvenienced. The police arrested two drone pilots that lived nearby but released them due to insufficient evidence.

THE WHITE HOUSE

The White House is off-limits to drone flying – period! In fact, all of the Washington, DC, is off-limits to drones. There is a no-fly zone within a 15-mile radius of Ronald Reagan Washington National Airport. The FAA wants it known to anyone that might consider flying a drone in the city that it is illegal and you will be prosecuted and fined to the fullest extent of the law … bottom line, don't fly anywhere near Washington, DC!

- Recent scare at the White House
  - In 2015, a man was arrested for flying a drone above Lafayette Park, which is across the street from the White House. The drone was visually spotted by the Secret Service and the pilot was arrested. As a side note, the radar that the Secret Service uses was not able to detect the drone.
  - In 2020, a drone almost crashed into Air Force 1 while landing at Andrews Air Force base. President Trump was returning from a trip when the drone just missed the aircraft. The pilot was never caught.

SPORTS ARENAS

The reason why sports arenas are out of bounds is because of the number of people they hold in one place. It would not be unheard of for a sports arena to hold 40–50 thousand people or more.

In September 2019, a drone pilot flew his drone over Michigan Stadium during a game. The Detroit police managed to identify the pilot and arrested two people in connection to the incident. According to the police, flying a drone over a stadium in Michigan is subject to arrest and/or civil penalties.

Flying near too many people is a bad idea for even the most experienced pilot. It's tempting to fly a drone to try and catch the action of a professional or college sports game but this would be a big mistake.

WILDLIFE REFUGES

Policy makers do not want drones near wildlife areas because some species might become frightened by the sight or sound of a drone. Animals might be driven into areas that put them at risk. You might think it's going to be difficult to catch someone in a forest flying a drone, but people do stupid things and post videos and pictures on social media and get caught.

In a study by National Geographic, scientists tested the reaction of drones engaged with black bears. Although the bears didn't look frightened, their heart rates increased significantly. The scientists had implanted heart sensors into the bears previously. Some bears ran with their cubs to avoid the drones while some just went back into their dens.

In 2020, an eagle in Michigan took down a drone and sent it diving into a lake, score one for the eagle. Most people that use drones for photographing wildlife have been photographing wildlife their whole careers with traditional camera equipment. They know when to leave an area or wildlife alone.

FOREST FIRES

According to the US Code of Federal Regulations, 43 CFR 9212.1(f), "it is illegal to interfere with the efforts of firefighters to extinguish a fire". This law was put in place to deter drone pilots from distracting firefighters or colliding with air-based firefighting equipment.

Not surprisingly, firefighters do use drones to help fight fires. Drones can go places that are just too dangerous for people to go into. They can give firefighters an aerial view as to how much the fire has spread, hot spots, and if personnel are in danger. Additionally, drones can fly in poor weather, carry high-resolution cameras, infrared sensors, and fly into places that conventional aircraft cannot.

MILITARY BASES AND DEPARTMENT OF ENERGY (DOE) SITES

- The FAA has banned the flying of drones over and around 145 military bases. The FAA sites national security concerns. Check the FAA site for the complete list.

The following are just a few DOE facilities and nuclear sites that are banned from drone flying:

- Hanford Site, Franklin County, WA
- Pantex Site, Panhandle, TX
- Los Alamos National Laboratory, Los Alamos, NM
- Idaho National Laboratory, Idaho Falls, ID
- Savannah River National Laboratory, Aiken, SC
- Y-12 National Security Site, Oak Ridge, TN
- Oak Ridge National Laboratory, Oak Ridge, TN

PRISONS AND OTHER CORRECTIONAL FACILITIES

Many states have implemented laws that ban a drone within so many feet of a correctional facility. This is in response to recent incidents where people from the outside have dropped drugs, money, and supplies to inmates on the inside of these facilities.

In 2018, the Virginia Department of Corrections reported 33 drone sightings near correction facilities. To site just one incident, a security staffer at the prison reported seeing a drone with a package attached to it on the side of the road. He reported it to his superiors. They called in the state police to investigate the incident. The results of the investigation concluded that the package contained drugs, money, a cellphone, and a handcuff key in the package.

People have gotten away with this because law enforcement cannot respond fast enough to these types of incidents. They do not have the technology to take down drones. The list of prisons that are banned from drone use is too large to list here. Check the FAA site for a complete list.

In October 2020, an ex-convict was arrested and charged with trying to use a drone to drop contraband into the Fort Dix federal prison. This included tobacco, cellphones, and chargers. The man allegedly dropped similar items in the prison in 2018.

NATIONAL PARKS

Authorities do not want drones in national parks because they could scare the wildlife, crash into animals and people, and ruin natural vistas. Park rangers now have the ability to fine drone pilots or confiscate them if needed. Many states now ban the use of drones in their park systems for similar reasons.

The FAA has banned the use of drones in the following landmarks and National Parks and Monuments:

- Statue of Liberty National Monument, New York, NY
- Boston National Historical Park (USS Constitution), Boston, MA
- Independence National Historical Park, Philadelphia, PA
- Folsom Dam, Folsom, CA
- Glen Canyon Dam, Lake Powell, AZ
- Grand Coulee Dam, Grand Coulee, WA
- Hoover Dam, Boulder City, NV
- Jefferson National Expansion Memorial, St. Louis, MO
- Mount Rushmore National Memorial, Keystone, SD
- Shasta Dam, Shasta Lake, CA
- Valley Forge National Park, Valley Forge, PA

In 2014, a tourist wanting to take pictures from a drone crashed it into the Grand Prismatic Spring. This followed many instances in national parks where drones have chased bison, big horn sheep, deer, and elk. Rangers are concerned about the effect that drones will have on wildlife.

SCHOOLS

Flying near a school is just a bad idea. The possibility exists that a drone pilot could crash into a group of school children. Flying a drone near school children with a drone capable of taking pictures is not a very good idea and should be avoided at all costs. No parent wants pictures taken of their children without their knowledge or consent.

At least four school districts in New Jersey have banned the use of drones over their schools. Ramapo Indian Hills, Fort Lee, Oakland Lake, and Franklin Lake prohibit remotely controlled Unmanned Aerial Vehicles (UAV's) flying on, over, or landing on school property while occupied during the school day by students, staff, parents, or community members.

CITY PARKS

Given any nice day, you will find people in a park. Wouldn't it be great to take your drone to a park on a beautiful summer day? Well, there is some good news here. The city of Raleigh, North Carolina is drafting a revised policy that would allow drone pilots to fly in most city parks. After months of deliberations, the city council decided to ease restrictions on drone use. Drones weighing less than 400 grams can be flown in city parks except those designated as wetlands or nature preserves. Sounds like a good reason to move to Raleigh!

LARGE SOCIAL GATHERINGS

Some states have banned the use of drones within so many feet of any large social gatherings. These include fairs, municipality celebrations, and feasts. The concern is that they may crash and do inadvertent harm to people or property. According to the FAA, it is illegal to fly your drone over crowds.

There have been many incidents where drones have crashed into crowds. A drone was involved in an incident in Ogaki, Gifu Prefecture, Japan where it crashed into a crowd injuring six people, including children.

On a beautiful Sunday afternoon, a drone flew into an MLB baseball park where the San Diego Padres were playing the Arizona Diamondbacks. The drone appeared to veer out of control and crashed into an empty seat. Due to the size of the drone and speed, it could have done some real damage to any spectator.

LOCAL LAWS

Check with your county or municipality. There may be additional local laws that apply to drones. All pilots need to be sure they are flying legally. Most local laws relating to drones usually ban drones from flying over city owned property or private property. Some of these laws may conflict with state or even federal laws. You can find out about any local ordinances by calling your city hall, township, or county and asking if there are any ordinances relating to drones.

As an example, in Ventnor, New Jersey the city prohibits drones from taking off or landing on any government or public buildings. Drone pilots cannot fly less than 400 feet around any of these buildings or operate in any city parks or city owned property.

Pennsylvania takes a broader approach. As of October 2018, no county, city, borough, or municipality in Pennsylvania is allowed to legislate the ownership or operation of a UAV. Any legislation is preempted by Pennsylvania state code Title 18 Section 3505 and by any FAA regulations.

DRONE RULES AND REGULATIONS

Although we will discuss rules and regulations in future chapters, the following is a good starting point for even experienced pilots:

- Keep your drone within line of sight or use a visual observer who is co-located that can spot it for you.
- Never fly near aircraft. Flying near aircraft, commercial or not, puts you and the aircraft in danger. If your drone should make contact with an aircraft, it could cause the aircraft to crash. The liability of a crash, lives lost, and damages will be your responsibility.
- Never fly near group of people. No matter how hard you try not to crash, the possibility always exists. Wind can blow your drone off course, a minor distraction, or malfunction (hardware or software) all could cause a drone to crash. Crashing into a crowd is a pilot's worst nightmare.
- Do not fly under the influence of drugs or alcohol. This is a good rule of thumb no matter what the sport. Flying a drone at high speed while intoxicated is a very bad idea and can put you and other people at risk. Reflexes are not as quick to respond and judgment tends to become impaired, making flying difficult.
- Do not fly after dark
- Ceiling limitation
  - Drones can only fly to a legal height of 400 feet Above Ground Level (AGL).

PART 107 PILOT'S LICENSE

Although not required for the average consumer, you can take a class and subsequent exam and become a certified drone pilot. This would allow you to fly your drone for commercial purposes. We will discuss why you might want to do this in future chapters.

- Pilot Classes
- License – Part 107 class allows you to fly your drone for commercial use
  - Videography
  - Photography
  - Construction inspection
- Commercial versus private
  - If you are going to use your drone for your personal use, you do not need a license or insurance. That means you cannot be hired to take pictures, videos, or use your drone for anything that involves being paid for your services.
  - If you intend to use your drone for commercial purposes, you need to take the Part 107 class. It's highly recommended that you get special drone insurance in case you crash your drone and cause damage or hurt someone.

REGISTER YOUR DRONE

Drones weighing more than 0.55 pounds and less than 55 pounds must be registered with the FAA. The FAA enacted this law in 2015. Although the FAA cannot force a drone owner to register their drone, failure to do so could result in a fine.

The FAA makes registration easy by providing a website at: https://www.faa.gov/uas/getting_started/register_drone/. The drone's owner must provide the appropriate information including name, address, phone number, etc. The cost to register a drone is $5.00 per drone and is valid for 3 years.

Commercial drone owners must register their drones under either Part 107 or Section 336. If an owner registers under Part 107, then it is being registered for recreational, commercial, or governmental purposes, and agrees to operate the drone under Part 107 regulations. Under Section 336, the pilot must fly their drone as part of an aero-modeling club and agrees to fly their drone under the Special Rule for Model Aircraft.

- Get a tail sticker
  - Once a drone is registered, the owner will receive a sticker in the mail with a registration number. The sticker must be placed on the drone so it can be easily seen. Stickers cannot be placed on unregistered drones under penalty or fine and every drone must have its own sticker.

- Drone registration database
  - The FAA takes the registration information and saves it in a database. The database is used for statistical purposes. However, if the drone pilot is involved in an incident, legal authorities can contact the FAA with the sticker number and request owner information.

DRONE ACCIDENTS

Accidents/incidents

- From August 2015 to January 2016, the FAA has received 583 complaints and reports on drone incidents. These incidents involve pilots flying into restricted airspace, buildings, and people.

In 2015, a drone fell into a crowd of people in Seattle Washington injuring two people. The pilot was found guilty of endangerment and received a 30-day jail sentence and a $500 fine.

In September 2016, a woman sued a University of Southern California fraternity and an event planner as the result of a drone crash, claiming head injuries.

In August 2016, guests at a wedding in New Hampshire sued the wedding party and an event planner after the groom crashed a drone and injured a woman, causing severe damage to her nose and orbital bone.

These were only the drone incidents that were reported. Also of note is that these were incidents that were reported as happening in the United States.

Reporting incidents

- Today, most people will report a drone incident to local authorities. Most police departments are ill-equipped to investigate incidents relating to drones. These types of investigations involve media forensic expertise, people that understand aviation forensics, and have a strong drone background. However, this is just a starting point. It's best to get the incident on the official record in case it leads to an arrest.
- Next, you need to report the incident to the FAA. The best place to report the incident to is the FAA Flight Standards District Office. You can find an office close to you by searching the FAA site. Every state has an office although they might not be close to your location.
- You will need several pieces of information to report an incident. This is a case where the more you have the better. Don't try approaching the pilot. Depending on the person, you might put yourself in jeopardy. While you are at the scene, take a photograph of the drone. You might

be able to zoom in on the tail number. If there is a car near the scene, take a picture of the license plate. Taking a video of the pilot breaking FAA regulations would be very helpful. Logging the date, time, and location is a necessity.

Government involvement

Recently, drones have been involved in collisions or close encounters with commercial aircraft. In these situations, the Federal Bureau of Investigation (FBI) and National Transportation Safety Board are brought in to investigate. If you witness a situation where a drone puts an aircraft in danger, call the police immediately and stay on site. Collect the information as described above.

DRONE INSURANCE

Like any other sport, it's a good idea to carry insurance. Insurance provides peace of mind should there be a problem when flying your drone. Insurance provides coverage for medical claims, property damage, and personal injury. Check with your current provider to see if they offer drone insurance. If you can add it to your current policy, it probably will be cheaper.

If your current provider does not provide drone insurance, it can be purchased from a multitude of places. Try to choose a company that has been in business for a while. Drone insurance has developed into a cottage industry with many people wanting to take your money.

Even if you are able to get insurance through your current provider, ensure you read the policy. You will need to know exactly what your policy covers. Here are some questions to keep in mind:

- Does the policy cover theft?
- Does it cover you when you are flying?
- Does it cover all of your drones or just a few or just one?
- How much personal damage does it cover?
- Does it cover medical expenses, when you hit someone with it?
- How much physical damage does it cover?
- Does it cover you, if your fly is for recreational and commercial use?
- Do I need it even though I use it for recreational use?

    - Even though pilots that fly drones commercially require insurance, a recreational user should still consider insurance. For the minimal amount a policy costs, it's well worth the investment. An incident where someone is injured could easily cost millions of dollars to settle, especially if it involved a child.

- What can I expect to pay for insurance?
  - Looking online, prices start at $7.00 (USD) per month for a basic policy for a recreational user and go up from there. A commercial policy starts around $500.00 (USD) a year for a basic policy and could be higher depending on the options you choose. Check with your insurance provider to get exact pricing.
  - Choose the insurance that makes sense for you. Your homeowners policy may cover you and provide the amount of insurance you need if you are just a recreational flyer taking out your drone occasionally. However, as a commercial flyer you would need to ensure you have yourself covered from all angles.

STATE DRONE LAWS

Almost all states have some sort of laws governing drones. Drone operators need to be aware of the laws that govern the states in which they fly. Failure to follow state laws could end up in fines or jail time for the drone operator. Not knowing is not an excuse. All laws are published and available online to everyone. More on drone regulations will be covered in future chapters.

PRIVACY

With the increase in drone usage comes the chance of misuse. Since even the most basic drone carries a camera, privacy is a major concern. It's very easy for a drone enthusiast to fly their drone a little bit above the fence line to see what is on the other side. Some states do have laws against flying over another person's property and when it involves the use of a camera, the laws are even stricter. Know your state laws before flying.

As an example in Florida, the state forbids using a drone to fly over another person's property. This is in violation of a reasonable expectation of privacy. In Arkansas, the state forbids drones from taking pictures over another person's property. This is considered video voyeurism. California forbids video recording another person without their permission. Nevada prohibits a weapon on a drone. Anyone who owns a drone should check with their state for laws regarding drones before flying them. The penalties, financial and jail time, could be rather stiff depending on the state.

Commercial drone pilots do not have any rights at all when it comes to the violation of privacy. In some states, commercial pilots cannot fly over your house and take pictures or videos without asking permission, and you have the right to say no.

Fortunately or unfortunately, the federal government is exempt from any of the above. As an example, the military does have the right to fly over your property. State and local authorities have more restrictions.

The real question here is how high above your property do you own? Well, it varies from state to state. Before the invention of air travel, landowners owned an infinite amount of space above their homes. This was known as, "Cujus est solum ejus usque ad coelum", which is translated from Latin into, "whose is the soil, his is up to the sky". As of today, the Supreme Court has not acknowledged an upper limit on this matter. However, the federal government considers anything above 500 feet as navigable airspace.

In the European Union (EU) privacy is taken much more seriously. Privacy is considered a basic human right. Any violation against it is considered illegal. Even unintentional collection of personal data (pictures, video, location, house numbers, etc.) that could easily happen with a drone could be a violation of basic human rights. The collection and storage of personal data is governed by the General Data Protection Regulation and is much more restricted in the EU than in the United States.

PRIVACY LEGAL CASES

In 2014, a drone pilot was arrested for flying a drone next to a hospital and taking pictures of patients in examination rooms. The pilot was arrested and charged with attempted unlawful surveillance. The pilot was found innocent. The jury cited that the operation of the drone did not violate the patient's privacy.

In 2015, a police officer in Georgia was found to be flying a drone over a neighbor's yard on several occasions. The officer was charged with felony eavesdropping and fired from his job. As we can see, the outcomes can be drastically different depending on the state and circumstances.

INTERNATIONAL LAWS

If you are living outside of the United States or you are planning on traveling with your drone, you must familiarize yourself with the respective countries drone laws. Failure to do so and you could wind up paying a significant fine and/or a jail sentence. If you are visiting another country, then you must obey all of the laws in that country. Being a visitor does not exempt you from their laws and not knowing is not an excuse. The following is a high-level breakdown of some of the laws you should be familiar with.

Europe

Countries in the EU, abide by laws as set out by the European Aviation Safety Agency (EASA). One big difference between Europe and the United States regarding drone registration is that in the United States the drone has to be registered with the FAA and in turn you receive a tail number. In Europe, the pilot must be registered with the Civil Aviation Authority. The pilot then receives a registration number.

EASA is still working to pass legislation on drone use. EU countries have agreed that drone use should fall into one of the three categories: Open, Specific, and Certified. Open use would equate to someone using their drone for recreational purposes. Specific use means that the drone could be used for riskier flights. Commercial use would fall into this category or drones that might come into contact with humans. Certified use means that the drone is involved in dangerous flights. This might include drones that someday carry people or dangerous cargo.

Austria

Drones that do not exceed 79 joules of energy or fly a maximum of 30 meters and weigh less than 250 grams do not require any approval from the government. The Austrian government does require you to hold $1 million (USD) dollars' worth of insurance in order to fly your drone. In addition, you cannot fly your drone in a city or town without a pilot's license. You can obtain a license from Austro Control. You cannot fly with First Person View (FPV) goggles unless you have a second person present to keep an eye on the drone. Finally, you cannot fly over airports, government facilities, or large crowds of people.

France

In France, the pilot must pass a test and be registered for all uses recreational or commercial. Drones may not be flown over 150 meters in uncontrolled airspace. You must have proof of insurance and an ID tag must be placed on your drone. The ID tag is good for 5 years. Drones may not be flown within 10 kilometers of an airport or over people. Currently, you cannot fly any drones over Paris without special permission from the government due to safety considerations.

Germany

Drone laws in Germany are determined by the German Federal Aviation Office. In Germany, there is no difference in laws between recreational and commercial use. All drone pilots who fly drones heavier than 2 kilograms, must take and pass a knowledge test. Any pilot flying a drone weighing less than 2 kilograms does not need any permission to fly. All drones weighing more than 0.25 kilograms must have a sticker on it displaying the owners name and address. A pilot wanting to fly a drone heavier than 5 kilograms must obtain special permission in order to fly it. FPV goggles can be used if you are flying below 30 meters and the drone weighs less than 0.25 kilograms. An exception would be if you have a second person spotting the drone.

Ireland

Drones are considered as model aircrafts, and the same rules apply to both the aircrafts. The governing body in Ireland is the Irish Aviation Authority (IAA). Drones cannot be flown over crowd of 12 or more. The drone must stay within 30 meters of the operator and be in a visual site. The drone cannot fly within 30 meters of a person or structure. All drones must stay below 120 meters, not fly within 5 kilometers of an airport, and are prohibited from flying near military installations.

Drones weighing more than 1 kilogram and less than 25 kilograms can be registered online. Any drone heavier than 25 kilograms must be registered in person with the IAA. Any drone heavier than this must be registered as a manned aircraft.

Italy

In Italy, drones are governed by the Italian Civil Aviation Authority (ENAC). Recreational drones must be flown within the visual sight of the operator. Drones cannot be flown over crowd or at night. Recreational drones cannot be flown higher than 70 meters, commercial drones cannot fly higher than 150 meters, and both are prohibited from flying within 5 kilometers of an airport.

Pilots wanting to fly drones for commercial use must obtain training, an operator's license, and a health certificate. Before every flight, the pilot must submit a statement of operations declaring exactly what they are planning to do with the drone including flight path information.

South and Central America

South and Central American countries have a few restrictions regarding drone use. There are some subtle differences, so check the laws for the specific country you are planning to fly your drone in. Some common regulations include registering your drone and the maximum altitude your drone can fly.

India

According to Indian laws, only Indian citizens may fly drones. As a foreigner, you may lease a drone from a registered agent. The agent must then file for a Unique Identification Number (UIN) for the time you will be flying the drone. Drones must be in the visual sight of the operator and fly below 400 feet.

Southeast Asia

For the most part, Southeast Asia welcomes drone use. There are 11 countries that make up this region: Indonesia, Philippines, Thailand, Cambodia,

Brunei, Timor-Leste, Laos, Malaysia, Singapore, Vietnam, and Burma. Brunei is the only country that has banned drones. If you take your drone to Brunei, it will be confiscated at customs and you run a risk that it will not be returned when you leave.

Most other countries do require a permit to fly. However, the process is quick and cheap to obtain one. Cambodia does not require anything to fly in the country whether it's for recreational or commercial use. Most countries in this area do not want you flying within 10 kilometers of an airport, so please take note.

Africa

Africa is a large continent and the rules vary as much as the countries do. Some countries do not allow flying drones and will confiscate your drone. Others have very relaxed regulations. You should check with the specific country before bringing your drone and risk getting it taken from you.

Oceania

Oceana is made up of 14 countries. Australia, Micronesia, Fiji, Kiribati, Marshall Islands, Nauru, New Zealand, Palau, Papua New Guinea, Samoa, Solomon Islands, Tonga, Tuvalu, and Vanuatu. Of all of the countries that make up this area, Australia and New Zealand have clear regulations on drones. In Australia, the Civil Aviation Safety Authority is the main governing body and in New Zealand, the Civil Aviation Authority of New Zealand is its main authority. As recommended before, check with the country you are planning on visiting to get specific laws on drones.

Countries to avoid

There are some countries where drones are banned or make it so difficult to get a license that it's just not worth the risk. In a few of the countries listed below, you could be arrested and jailed for flying in their country.

Algeria, Azerbaijan, Antarctica, Bahrain, Belize, Bhutan, Brunei, Cuba, Egypt, Iran, Iraq, Libya, Madagascar, Morocco, Nicaragua, North Korea, Oman, Sri Lanka, Syria, Tunisia, Uzbekistan, and Vatican City

SUMMARY

This chapter is an important chapter. Regardless of how good a pilot you are and how long you have been flying, you must follow any government issued laws, state, and even local laws. The single most important fact you can take away from this chapter is there are many laws regulating drones today. Know where you can and cannot fly your drone!

Chapter 4 discusses drone hardware and software. In this chapter, we discuss hardware and software manufacturers and associated costs for purchasing a drone. Consumer, commercial, and military drones are discussed and compared. Of great importance is a troubleshooting section. Should you have a problem with the motors, controller, or camera, you will learn techniques to check if each is operating properly.

Chapter 4

Drone hardware/software

Image used from the unsplash: https://unsplash.com/photos/yFnX8DaC3UM
Drone Hardware

Drones generally fall into one of the three categories: consumer, commercial, or military. We will compare consumer drone capabilities with commercial drones here, since military drones are really in a class by themselves and will be discussed later in the chapter.

CONSUMER VERSUS COMMERCIAL DRONES

- Extensibility
  - Commercial drones can be modified to carry different payload configurations from Light Detection and Ranging (LIDAR) to multiple cameras. It's possible but more difficult to modify

consumer drones to carry similar payloads. We will discuss how to do this in a later chapter.

- Capability
    - Commercial drones can be used for a multitude of uses including searching for people, delivering packages, and dropping life preservers to people drowning. Consumer drones are not made to do any of the above but can be if you have the ingenuity.
- Speed
    - Consumer drones excel in speed. The Drone Racing League uses modified consumer drones to race with. Commercial drones do not need to have this speed capability. They are generally the "working drones" and speed is usually not a necessity.
- Payload weight
    - Commercial drones generally can lift much more than consumer drones. This is what commercial drones are designed for. Consumer drones do not need to lift heavy loads. This is not to say that they can't, but their payload is usually limited.
- Price
    - Given the above, I don't think it's a surprise to see that commercial drones can cost a lot more than consumer drones. An entry level consumer drone can cost as little as $20–30 (US), while an entry level commercial drone could easily cost over $5000 (US).

CONSUMER DRONE MANUFACTURERS

- Today, there are many companies that manufacture drones. The following is a shortlist of some of the more popular consumer manufacturers and selected models:
    - DJI
    - Parrot
    - Yuneec

Image used with permission from Ralph M. DeFrangesco
DeFrangesco, R. (2021). Phantom professional DJI 3. Williamstown, NJ

Image used with permission from Ralph M. DeFrangesco
DeFrangesco, R. (2021). Protonic Drone. Williamstown, NJ

Image used with permission from Ralph M. DeFrangesco

DeFrangesco, R. (2021). Parrot Drone. Williamstown, NJ

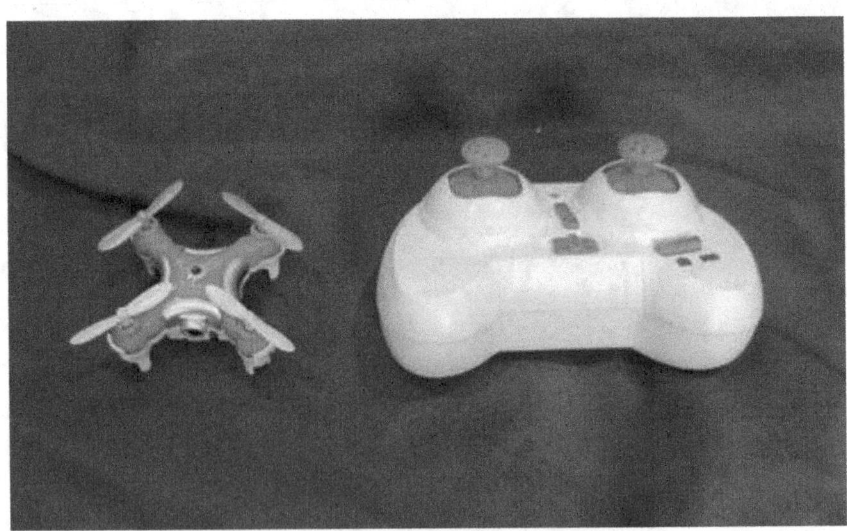

Image used with permission from Ralph M. DeFrangesco

DeFrangesco, R. (2021). Micro Mini Drone. Williamstown, NJ

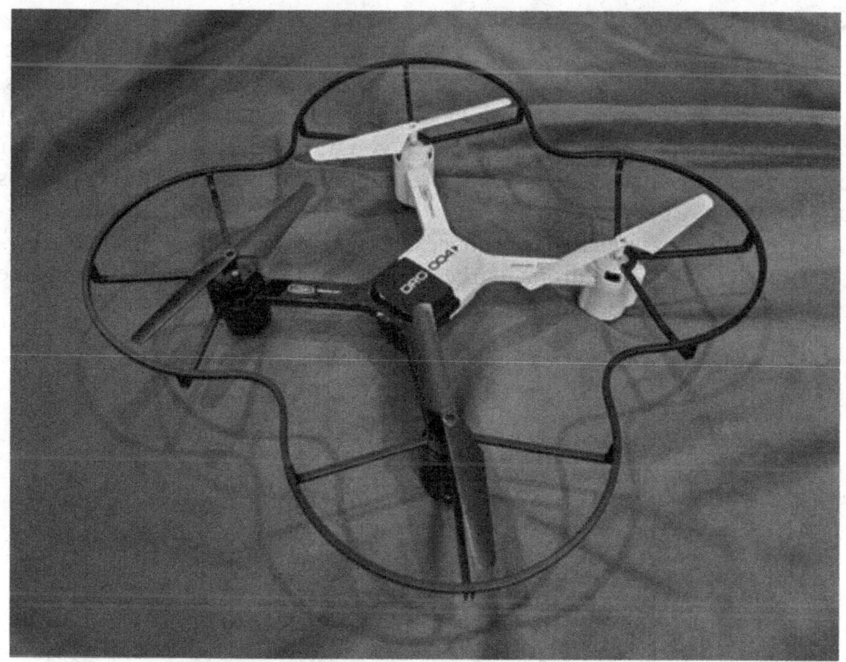

Image used with permission from Ralph M. DeFrangesco
DeFrangesco, R. (2021). Sharper Image Drone. Williamstown, NJ

FIXED-WING DRONES

Fixed-wing consumer drones are fairly new. This type of drone works differently than a quadcopter. Typically, they only have two motors. Rather than taking off from the ground, they are thrown into the air. It is possible for one person to fly the drone, but works best with two people.

The controller is similar in that it looks like a quadcopter controller, but it controls the motors and ailerons, which cause the drone to go up and down along with motor speed. Cheaper models are usually made of styrofoam while more expensive models could be made of plastic or carbon fiber, which would increase the price significantly.

Fixed-wing drones fly similar to a quadcopter. As mentioned above the takeoff is different and the landing seems more like a controller crash. They do come with landing gear, but you would need a perfect piece of ground to land it on its wheels, like the way a plane lands.

UNDERWATER DRONES

Underwater drones are the newest class of drones available to consumers. Underwater drones tend to be very expensive and you could imagine why. It takes a lot to make a drone waterproof. If you want to learn how to build your own underwater drone, or just want to understand how they work, there is much more in Chapter 8.

Many drone manufacturers produce a myriad of drones. It seems like everyone is cashing in on the drone market today. The following is a chart of some of the more popular drones with capabilities and cost estimates.

Manufacturer	*Model*	*Range (kilometers)*	*Camera resolution*	*Flight time (minutes)*	*Cost (USD)*
DJI	Air 2S	8	20MP	31	$999
DJI	Mavic Air 2	10	12MP	34	$799
DJI	Mini 2	10	12MP	31	$499
DJI	Mavic 2 Pro	8	20MP	31	$999
DJI	Mavic 2 Zoom	8	12MP	30	$1329
Ryze	Tello	100 meters	5MP	13	$99
DJI	Phantom 4 Pro	8	20MP	30	$999
PowerVision	PowerEgg X	6	12MP	30	$1245
DJI	FPV	10	12MP	20	$739
Parrot	Anafi	2.4	21MP	25	$691
DJI	Inspire 2	7	30MP	25	$2599
Parrot	Bebop	2	14MP	25	$775
Syma	X5HC	100 meters	2	9	$66
UDI	818A-HD+	Unknown	2	8	$60
Potensic	D80	Unknown	720P	20	$200
Holy Stone	HS165	300 meters	1080P	15	$150

Although these are not in any order, DJI certainly does top the list. You can see that the majority are close in their flight time capability, but there is a significant gap in their range in which they can cover the camera resolution varies and, of course, the price.

COMMERCIAL DRONE MANUFACTURERS

Commercial drones look similar to their consumer counterparts but vary in capability. They are designed to carry bigger payloads, have longer-lasting batteries, and fly further. The cost of the drone typically increases when additional capabilities are being added. The following are a few popular commercial drone manufacturers and models.

Image used with permission from RMUS
Police-fire-search-and-rescue-drones. (2021). Retrieved from https://www.rmus.com/collections/police-fire-search-and-rescue-drones
RMUS Law Enforcement Drone

The above drone solution is offered by RMUS. It includes a 640 × 512 radiometric thermal camera and daytime 10× Red, Green, Blue (RGB) zoom camera that are remotely controlled. The drone includes a Watts Prism Quad aircraft, RMUS position lighting system, and a ground control station with charger.

Image used with permission from DraganFlyer
Quadcopters-multirotors. (2021). Retrieved from https://draganfly.com: https://draganfly.com/products/quadcopters-multirotors/
DraganFlyer Commander Rededge MX

Image used with permission from DraganFlyer
Draganflyer-Commander-v1.2.pdf. (2021). Retrieved from https://draganfly.com: https://draganfly.com/wp-content/uploads/2020/04/Draganflyer-Commander-v1.2.pdf
DraganFlyer Commander-QX100

Image used with permission from DraganFlyer
Quadcopters-multirotors. (2021). Retrieved from https://draganfly.com: https://draganfly.com/products/quadcopters-multirotor/
DraganFlyer M600 Pro

The following table is a list of top commercial drones. It is not unusual to see consumer drones on this list. Many professionals who fly drones like the way the drones handle, their capabilities, and most of all, their price.

Manufacturer	*Model*	*Range (kilometers)*	*Camera resolution*	*Flight time (minutes)*	*Cost (USD)*
DJI	Matrice 100	2	N/A	40	$3000
DJI	T600 Inspire	6	12MP	20	$3000
DJI	Phantom 4 Pro	7	20MP	25	$750
Hubsan	H301S Spy Hawk	1	1080P	30	$360
Vantage	Vesper	45	1080P/ Thermal	60	$8000
Autel	EVO II	9	48MP	40	$1200

The brand DJI is one of the more popular selling drones that consumers have recognized due to its quality and affordability. Just like the consumer drone list, DJI is a top-selling brand. However, many commercial drone manufacturers make quality drones. Similar to any investment or purchase, this will depend on your requirements and your price range.

MILITARY DRONES

Military drone capabilities will differ widely from consumer and commercial drones. Military drones have two functions: reconnaissance and surveillance and can be utilized as an attack platform.

Taken from the public domain: https://en.wikipedia.org/wiki/General_Atomics_MQ-1_Predator
MQ1-Predator

The MQ-1 Predator has a long and dependable history in the United States drone arsenal. General Atomics Aeronautical Systems first produced the Predator in the early 1990s as the RQ-1 Predator. The initial mission of the

Predator was for reconnaissance purposes. It was renamed the MQ-1 because of its multi-purpose role, including its attack capability. In 2000, the MQ-1 was fitted with Hellfire missiles and successfully attacked a tank making it a perfect attack platform. The Predator was successfully used in Bosnia, Kosovo, Syria, Iran, and Afghanistan. In 2018, the US Airforce retired the Predator and replaced it with the MQ-9 Reaper. The Predator pictured above is flying over a desert and is configured with a Hellfire missile.

Taken from the public domain: https://en.wikipedia.org/wiki/General_Atomics_MQ-9_Reaper
MQ-9 Reaper

The MQ-9 Reaper is developed by General Atomics Aeronautical Systems for the US Airforce. The Reaper is a hunter-killer Unmanned Aerial Vehicle (UAV). The platform caries a variety of weapons including the GBU-12 laser guided bomb, AGM-114 Hellfire missile, the AIM-9 Sidewinder missile, and the GBU-38 Joint Direct Attack Munition. The Reaper pictured above is shown flying over a desert.

Taken from the public domain: https://en.wikipedia.org/wiki/General_Atomics_MQ-1C_Gray_Eagle
Army's Gray Eagle

The MQ-1C Gray Eagle is also known as the Warrior, Sky Warrior, and Extended Range Multi-Purpose UAV. General Atomics Aeronautical Systems develop the platform for the US Army. The UAV was first developed in 2002. It entered service in 2009 as an attack platform. It is capable of carrying a Hellfire missile or GBU-44/B Viper Strike guided bomb. The Gray Eagle is still deployed today, supporting troops in South Korea and Niger. The Gray Eagle pictured above is shown flying over an unpopulated area.

DRONE COSTS

Drones vary widely in price. Small consumer toy drones can be purchased for under \$20–30 (US). These drones have minimal capabilities and features. On the other side of the spectrum, DJI makes some of the high-end drones that can easily top a thousand dollars but can include many more capabilities than the under \$30 drones.

Commercial drones can be much more expensive than consumer drones. Commercial drones can run into thousands of dollars depending on the functionality. However, we should note that to use a drone for commercial purposes, which entails making money with the drone, you need a Part 107 license. We will discuss how to obtain the license and insurance in subsequent chapters.

Military drones could cost millions of dollars. Cost depends on their purpose and, of course, their configuration. High-definition cameras are expensive, and the ordinance is even more expensive. These drones are only flown by military personnel. A civilian cannot drop an explosive device or fire a missile on another country.

Additionally, Military drones can be flown by pilots located thousands of miles away and are typically flown in teams. One team member is the pilot, and the other operates the camera or manages the ordinance and acts as a backup pilot if needed.

CONSUMER DRONE ADD-ONS AND ACCESSORIES

Sensors

There are many sensors available to drone enthusiasts today. However, sensors are limited to the drone type or manufacturer. Most add-on sensors are for commercial use. These can include LIDAR, ultrasonic sensors for distance, thermal sensors, chemical sensors, and depth cameras.

- LIDAR sensors use ultraviolet light to image objects or provide ranging capabilities. LIDAR is used in many cases, including accident scenes, forestry, agriculture and landscaping, terrain modeling, and archeology. A LIDAR system for ranging can be purchased for as little as $100.
- Oxygen sensors can measure the amount of oxygen in the air. These sensors can be mounted to drones and flown into areas that a human should not enter into. An oxygen sensor can be purchased for under $100.
- A thermal sensor can detect human movement and can be purchased for under $100.
- Forward-Looking InfraRed (FLIR) cameras can be an excellent addition to any drone. Hobbyists can purchase a FLIR camera for their drone for under $500. FLIR has an additional capability of connecting to smartphone devices for under $300.
- Several cellphones have time-of-flight (ToF) or depth cameras as standard equipment. For example, the following cellphones have ToF capability, the Samsung Galaxy S10 5G Ultimate Edition, LG G8 ThinQ, and iPhone 2020. Thus, allowing the average hobbyist 3D imaging capability. With a bit of ingenuity, this could be mounted to your drone.

- Mission planning software

  - Mission Planning Software (MPS) allows a drone pilot to layout and test their flight without actually flying a drone. You can set camera angles, speed, flight path, and even trigger sensors all in a virtual environment. MPS software can run on a computer or even a mobile phone.

- Cameras
  - Almost all drones have some picture or video capturing capability. The number and quality of cameras vary by drone manufacturer and line of drones. For example, a top drone manufacturer, DJI, offers a 4k camera mounted on an active gimbal. An active gimbal allows the drone operator to lock onto a target and follow the object with smooth camera motion and limited jitter. Smaller drones have picture capturing capabilities but usually have cameras with fewer pixels.
- Global Position Systems (GPS)
  - GPS capabilities are used on higher-end drones. Drones that have a GPS capability will fly based on utilizing GPS coordinates. Thus, a drone can be sent to a specific GPS coordinate or programmed to fly to a location using GPS coordinates.
- Recording software
  - Software that is available for flight planning allows the user to record a flight path and utilize the exact flight path for replay later. Pilots who race drones might find this valuable to see how well they are flying, but commercial pilots might use it to see the area they cover using a scanning sensor.
- Controller
  - Drones vary in how they are controlled. Some drones have a dedicated controller, while others use a cellphone, touchpads, or a MacBook. Some drones use a combination of each device. The drone is flown with the dedicated controller, but the pictures and video come back to the cellphone or MacBook.
- Drone pontoons
  - Small and lightweight, pontoons add another dimension to your drone. With these accessories you can land or takeoff from the water. If you want to take nature pictures, you will find these handy.

- Propellers
  - Changing your propellers out can change the dynamics of your drone.
    - Larger propellers will give your drone more power and lift. Generally, larger propellers add stability. You may notice that your drone is a bit less responsive when you fly with larger blades. This is normal and the price you pay for stability.
    - Smaller propellers will make your drone fly faster. Your propellers will spin faster, so the drone will fly faster. However, you do lose stability when you go with smaller propellers.
    - There is no fast rule for deciding what size propellers to use. You can research to see what other drone pilots are using or you can experiment. Propellers are relatively cheap unless you purchase the carbon fiber type.
- Batteries
    - You more than likely have the standard battery that came with your drone. If you want to make your drone more powerful and fly longer, look to upgrading them to a battery that has a higher mAh (Milliamp Hour). Check with the manufacturer for the maximum mAh that you can use in your drone, or you will burn out the electronics.
    - Make sure that the voltage is the same as the original battery or this will cause problems. You cannot just increase the voltage of the battery.
    - Watch the weight of the new battery. If the weight is significantly higher, then you are defeating the purpose of putting in stronger batteries.
    - Generally, stronger batteries will allow your drone to lift heavier payloads.
    - Today, wireless charging a battery is possible. However, you need to make sure your drone battery can be wirelessly charged. If not, this will cause damage to your battery. If your drone battery can be charged wirelessly, then this opens a whole new door for you.
    - Chapter 9 discusses Do-It-Yourself projects. As an example, one project shows how to use your drone to watch your house. In order for your drone to be ready when you need it, the battery needs to be fully charged. One way to ensure it is always ready is to use wireless charging. One caveat is that the drone needs to fit over the wireless charger with the battery in it. Not all drones will be able to take advantage of this option. You will have to experiment to see if your drone will fit over the charger.

- How to choose a drone for you
  - Now that we have a good idea as to what drones and accessories are available, let's end the chapter by discussing the factors in choosing a drone.
    - For consumer use, the factors are few in number. The main questions are:
      - What will you use it for?
        - Entertainment, racing, etc.
      - What is your budget?
    - For commercial use, the factors vary a bit more:
    - What will you use it for?
    - What is your budget?
    - What accessories or sensors will you mount on it?
    - Do you need mapping or GPS software?
    - What kind of flight time do you need?
      - Related to battery life

TROUBLESHOOTING

Troubleshooting a drone is part art and part science. There are a myriad of techniques and software to troubleshoot drones, so we will only discuss a few, but these will apply to most drones.

Let's start with the basics:

- Are the batteries in the controller good?
- Is the battery in the drone fully charged?
- Do all propellers spin freely?
- Do the lights on the drone light up properly?
- Can you download the operational logs from the drone? (See the File Transfer Protocol (FTP) section to do this)
- How do you know if your controller is transmitting, or your drone is just not receiving the signal? We can quickly test to see if your controller is transmitting by capturing the controller's signal.

You will need the following hardware and software:

- Laptop (running Windows)

- Antenna such as HackRF One, BladeRF, RTL-SDR, LimeSDR
- Universal Radio Hacker if you are using Windows

Use the following procedures (Windows):

1. Connect the antenna to the laptop and boot it up
2. Start Universal Radio Hacker
3. Select Spectrum Analyzer
4. Select the antenna from the device pulldown list
5. Select the frequency that your controller is transmitting on (you can get this from the manufacturer or sometimes it's on the back of the controller)
6. Set the gain at 0, IF Gain at 32 and the baseband gain at 34
7. Leave the DC correction box checked and the Bias Tee unchecked
8. Click start

Do not turn your drone on at this point. You should see an Radio Frequency (RF) signal in the readout pane. Put your controller near your antenna and start pressing buttons or moving the joy stick. You should see the signal increase in the waterfall chart at the bottom or an increase in the RF signal in the top box. Either one indicates that the controller is working. Now, it might not be sending the correct signal, but it is sending a signal.

Testing with Linux

You will need the following hardware and software:

- Laptop (running Linux)
- Antenna such as HackRF One, BladeRF, RTL-SDR, LimeSDR
- GQRX

Use the following procedures if you are running Linux:

1. Plug in an antenna and turn on the laptop
2. Start GQRX
3. You will have to select your antenna and speaker output
4. Start the capture
5. Look for a signal at or near the frequency your controller is transmitting at

What if one or more motors are not spinning?

Let's try something simple first. Try taking some compressed air and blow out the motors. Sometimes dirt and other particulate get wrapped around the motor spindle. If this does not work, check the connection to the motor and controller board. It might be possible that the motors become disconnected.

If neither of the above work, then let's check if the motor is getting power. You will need a multimeter to check voltage. Some drones have the connections to their motors exposed. If you can see the connections to the motors, then place both probes on the motor connections. It does not matter which probe goes on the motor connections. Switch the multimeter to DC voltage. More than likely, the voltage will be around 3 Volts DC, but check the specifications from the manufacturer. Turn on the drone and the controller. Hold the drone and move the throttle up. You should get a voltage reading on the meter.

If your drone's motors are not exposed, on some drones you can see the board that the motors are connected to. Touch both probes to the board where the motors are connected and again turn on the drone and throttle up the motors. You should get a voltage reading.

If your drone does not have motors exposed or you cannot see where they are connected to the controller board, you will have to take the drone apart. You should see the motors at this point. You may even have to cut into the wires that connect to the motors. If you have to go this route, you can get some heat-shrink wrap to repair it.

What if your camera does not work?

This is a common problem with drones, since the camera is exposed and susceptible to damage due to crashes or exposure to the elements. Look at the camera for obvious damage including the wires that connect it to the drone. The drone is powered by the battery. You will need a multimeter to check the voltage to it. Using the same technique as above, place the meter's probes to the connections to the camera or where it connects to the drone. You should see approximately 3 Volts DC but check the specifications from the manufacturer.

Analyzing the controller signal

There are times when you will need to analyze the signal coming from the controller to the drone. Using the example above, the controller may be transmitting, but how do you know if it's transmitting the proper signal?

You will need the following hardware and software:

- Laptop (running Windows)
- Antenna such as HackRF One, BladeRF, RTL-SDR, LimeSDR
- Universal Radio Hacker

Use the following procedures (Windows):

1. Connect the antenna to the laptop and boot it up
2. Start Universal Radio Hacker

3. Select Spectrum Analyzer
4. Select the antenna from the device pulldown list
5. Select the frequency that your controller is transmitting on
6. Set the gain at 0, IF Gain at 32, and the baseband gain at 34
7. Leave the DC correction box checked and the Bias Tee unchecked
8. Click start

When you click on the controller, you should see a spike in the RF readout in the top box and a spike in the waterfall box at the bottom. You will need to move the pointer at the top to adjust to the exact frequency your controller is transmitting at. Click on stop. Go back to the Main panel and click record signal. The file will be pretty big, so only record a few seconds of the signal and save it out to a file using the save icon. If you change the default name, you will have to use the ".cs8" file extension.

Now that you have the signal captured, change the modulation to "PSK". The signal view should be analog and the show signal box checked with the pulldown "Bits". The box to the right will change to show you a binary representation of the signal.

You will have to get the manufacturers specifications to compare your signal to. This could be rather hard, but you might be able to find signals that other drone pilots have posted online or you can send the file to the manufacturer for analysis.

Other troubleshooting

Not everyone will have a quadcopter. Some people might have a Tera-firma vehicle, an underwater drone or a fixed-wing aircraft. The above trouble shooting suggestions will work no matter what type of drone you have. You will have to use your head a little bit on how to apply these techniques. Motors are motors and require voltage to work. You just need to know how to check it.

SUMMARY

In this chapter, we discussed drone hardware and software. We presented consumer, commercial, and military drones, and compared consumer drones against commercial drones, hardware add-ins, costs associated with purchasing them, and a handy section on troubleshooting.

Chapter 5 is an exciting chapter, since it discusses how to fly a drone. We start out reviewing safety tips, which should be a priority for all drone pilots. Registration of your drone should also be a top priority for all responsible drone pilots. A review of registering your drone, where to go, and how much it costs will also be reviewed.

Do you know the best place to fly? Do you know if your drone has auto takeoff/land, obstacle avoidance, or where the emergency land button is? Do you know what First Person View (FPV) is? We will discuss all of these topics at the end of the chapter, and we will review some memorable drone incidents that have made the news and social media broadcasts.

Chapter 5

Flying a drone

Image used from the public domain: https://commons.wikimedia.org/wiki/File:Drone_pilot.svg
DronePilot

This chapter was a fun chapter to work on since this is where the rubber hits the road or is it where the plastic hits the sky? Some people can fly drones, and some people can "really" fly drones. Some people have the natural ability to do tricks and do some exciting things while flying their drones. This chapter provides some safety tips before flying your drone. A fundamental process you must follow before flying is to register your drone, and we walk you through the process of doing that. There are certain places you can and cannot fly. We discuss this multiple times in this book, and this is just another reminder to know before you fly.

Even expert pilots crash their drones. A prepared pilot would want to know what parts you should have on hand. We suggest a few spare parts to carry. You may want to fly using First Person View (FPV) goggles. We discuss what is on the market and how you might fly using FPV. Today's drones have many

DOI: 10.1201/9781003201533-5

features such as obstacle avoidance, return home, Autoland/takeoff, and some drones can perform stunts with the push of a button.

Since we are talking about flying your drone, we want to take the opportunity again to discuss some notable drone incidents. A theme we are trying to convey is that flying a drone should be taken seriously and what better way than to share stories of what could go wrong.

Spatial disorientation is a serious problem that some people suffer from. We discuss what it is and why people suffer from it and how it differs from visual perception problems.

SAFETY TIPS

- Safety tips are essential and should be reviewed before flying the drone as it ensures a safer flying experience, which also adds to the enjoyment of flying. Before flying, check the drone for damage and ensure the battery is at total capacity to measure the approximate flight time. Know the drone and the functionality of the controls. If there is a home button for an emergency landing, ensure it has been configured to the home setting before takeoff. Proper configurations will ensure the drone has an emergency route back to the operator.
- Safety tips before flying
  - August 29, 2016, the Federal Aviation Administration (FAA) implemented a flying restriction that identifies where you can't fly your drone to maintain a safe airspace for not only yourself but others in your flying area.
  - Know your surroundings
    - Powerlines, wires, trees, cars, people, rivers/lakes, children play areas, local airports, walking paths, parks, etc.
  - Fly your drone in the line of sight (LoS). This means in your LoS, if you can't visibly see your drone then you are not in the LoS.
  - Do not fly near other aircraft.
  - Do not fly over groups of people, public events, or stadiums full of people.
  - Do not fly near or over emergency response efforts.
  - You should never fly above people, animals, or moving vehicles.
  - Do not fly near or interfere with manned aircraft operations.
  - Do not fly after dark.
  - Many outdoor areas such as parks, ski resorts, community parks, and other outdoor sporting areas DO NOT permit unmanned aerial vehicles (UAV). Fines, penalties, and even the confiscation of your UAV by an authorizing official could be a hefty consequence. We will discuss this in depth in other chapters.

- Check the weather and avoid flying when it is windy. Drones are unstable in windy conditions, and rain could compromise all functionality and could cause severe damage not just to your UAV but also surrounding objects due to failed components. The home function on the specific drone might not take wind conditions into consideration when returning home and could prohibit or delay the home function from completing and returning the drone to the expected location.
- Registration
  - How to register
    - The best place to get started is to register your drone in the FAA. The following website is where to register your drone:
    - https://www.faa.gov/uas/getting_started/register_drone/

In order to register your drone, you will need to provide some basic information:

- Email address, physical address, make and model of your drone, and a credit/debit card.
- The registration fee is $5 (US) per drone and the registration is valid for 3 years.
- In order to register your drone, you must meet the following requirements:
  - 13 years of age or older
  - A US citizen or permanent resident
  - Foreign applicants will be issued a certificate that is a recognition of ownership rather than a US aircraft registration

When the registration arrives, the certificate must be on the operator at all times when flying the drone. Any Federal, State, or local law enforcement officer can request to see your certificate.

Failure to register a drone that requires registration can result in a fine or imprisonment. The FAA can assess a fine of up to $27,500. Criminal fines can be assessed as high as $250,000 and/or 3 years in jail. For five dollars, do not be the poster child for not registering your drone.

- Label your drone
  - After you get your registration number from the FAA, you must affix a label with your registration number to the drone. You can print or handwrite the number to a label as long as it is legible. The label should be in a place that is viewable.

- You are now responsible
    - Keep in mind that if you use your drone for something illegal or cause damage with your drone, the label ties the drone to you directly, so fly responsibly!

- Know your drone
  - Controls
    - Distance – what happens when you hit the maximum distance that your controller works with your drone? Does your drone auto return home? Does it auto land?
    - In flight time – you need to know the approximate time that you can fly your drone with some added time to return back to you. Some drones have a battery charge indicator right on the display and others do not. If your drone does not, then it's even more important to know your drone's limits.
    - Lighting identification – does your drone have a separate control for lighting? This will help to identify your drone in low lighting conditions.
    - Sound identification – this might seem superfluous, but I am always listening to the sounds my drone makes. Is it straining? Are there any issues with the propellers? It's like listening to a car engine, you can tell a lot from the sound it makes.
    - Controller buttons – what do the buttons do? Do you know how to go forward, backward, or hold the drone in a hover?
    - Camera – Do you know how to turn on the camera to take a picture? Do you know how to take videos?
  - Modes
    - Manual Mode – most pilots fly in manual mode. This gives you more tactile control over the drone.
    - Altitude Mode – this will vary by manufacturer, but the basic idea is that you set the maximum altitude your drone can fly.
    - Auto/waypoint programming – you would use this mode in conjunction with navigation software. You can preprogram in a course and the drone will follow it using way points.
    - Normal Orientation Mode – this will vary by manufacturer, but it resets the drone to manual mode and points the camera in the direction you are flying.
    - Return to home (Fail safe) – This is a built in safety system for your drone. Let's say you are 100-yards out and over a heavily wooded area. Your battery is about to die. Using return to home would bring the drone back to you before the battery dies and

you lose the drone in the wooded area. You find this capability mainly on more expensive models.

- o Takeoff – almost all drones, cheap and expensive, have a takeoff capability with the push of a button.

LOCATION

As the saying goes, location, location, location! Flying location does make a difference. Even after you have done all of your checks, you could still be in a bad place to fly your drone. There have been many reported instances where drone pilots have lost control of their drones due to flying near high-voltage lines or where a lot of metal is in the area such as a factory.

Areas to avoid that might interfere with your drone are radio transmission towers, mining areas, high-voltage lines, substations, cell towers, microwave towers, train tracks, and high fences. Some pilots have noted that flying over or near parking garages and junkyards has caused problems, probably due to the high metal concentration.

The point here is that if you notice anything strange with the response in your controls, land your drone and move to a different location. It's not worth taking the chance that you might cause a crash or lose your drone.

EMERGENCY LAND – it might not seem obvious, but if you have never flown a drone before, or are flying a new drone, it's very easy to lose control of a drone. We knew a woman that bought her son an expensive drone for Christmas one year. She never flew a drone before but decided to try and teach her son how to fly it. The drone must have had a preprogramed flight path in it already, so when the drone took off; it disappeared into the sunset, never to be seen again. This is a tough lesson to learn for a first-time pilot with an expensive drone.

The proper thing to do when flying a drone you have never flown before is to know the controls, especially the emergency land, panic button, or kill switch, whatever the drone manufacturer calls it. This is just good practice in general because you could lose control of the drone and crash it into a building or person. Knowing how to land the drone safely before it does damage or gets away from you will save you grief and perhaps a jail sentence or fine in the future.

- Drone Bag (repair kit)

When flying the drone, you might crash, and it might need repairs. Your drone bag should include a small Philips and flat head screwdriver, needle nose pliers, a spare battery for the drone, and propellers. You should also include additional hardware needed to install the propellers and spare batteries for your controller. In most cases, higher-end drones will include tools to do this kind of work.

- Drone maintenance

Your drone should require little maintenance. I will caveat that statement with it depends on where you fly and how you fly. The motors do not require any oil so no need to even touch them. However, if you take your drone to the beach for the day and remember to check before you take it there as to whether or not you can fly there, when you are finished flying for the day, take a can of compressed air and blow out the motors, where the propellers meet the motors, the camera, gimbal, and in the battery compartment.

The batteries for the drone should be stored in a bag approved for battery storage. Unless the drone is being used every day, the batteries should not be kept in the charging unit. Run the batteries down, store them and only charge them when needed. The batteries in the controller should also be removed when the drone is not in use. I have seen many instances where the batteries have broken open and caused havoc on the electronics due to inactivity of a device. This is especially important during the hot months or if the batteries are stored near a heat source.

Depending on where you fly your drone or if you crash it like I have many times, you may need to clean it every once in a while. Use a light plastic cleanser and ensure not to get any cleaner on the registration sticker or drone markings. Some cleaning solutions will cause the stickers to come off. Obviously, if you have crashed your drone, then you will want to make sure all dirt and grass has been removed. Even a little dirt could cause the drone to be out of balance. Regularly cleaning the drone will ensure that the motors do not work harder to keep the drone balanced.

Finally, depending on the camera, it will determine how often maintenance is needed. The camera on the drone should be treated just like any other camera; it should be handled with care. A camera has a glass or plastic lens that can become dirty from flying, which a typical camera is not regularly exposed to.

Bugs are the biggest enemy. Before flying the drone, clean the camera lens with an approved lens cleaner recommended by the manufacturer. It is good practice to clean the lens after every flight as well. Regular cleaning will ensure the removal of any dust that it might collect between any flights. Use a rag that will not cause scratch marks.

SPARE PARTS

You will find that with many drone purchases, you are provided with spare parts for your UAV, including propellers. You will need to know how to remove and install new propellers as needed, especially if you are out flying and need to do an on the spot repair.

- Labeled Identification:

  - Your propellers can be identified in a variety of ways. The UAV should be equipped with two sets of two different props: one set of "clockwise" (CW) and one set of "counter-clockwise" (CCW).
  - When labeled like so, the "R" stands for CCW while the lack of an R means the prop is a CW prop.
  - Propeller identification: Look at the propellers on the drone; the propellers should be identified as A & B or 1 & 2 or R & non-labeled, which indicates CW. In some cases simply identifying the propellers are as easy as identifying which direction they are spinning. Propellers match the opposing sides diagonally. The installation and removal will also spin in opposing directions.

- Non-Labeled Identification:

  - Perhaps a better identification that is universal would be to identify the leading edge, which identifies the direction the propellers spin.

    The leading edge is identified as the edge that is raised compared to the other side of the propeller. This is identified with a slight concave curve underneath it. All propellers have a slated side, and identifying this early will help with the installation of new propellers as needed.

- Calibration:

  Calibrating a drone is essential, especially if you notice it is not flying and responding as expected. In many cases, when relocating a drone to a new flying area, calibration is needed to reset the compass location. Altitude drops, unstable drifts, or unstable hoovering are a good indication that calibration is needed. After any impact, calibration is essential before the next flight.
- For a successful calibration, ensure the drone is clear from surrounding metal objects and objects that could interfere, such as nearby cellphone towers and powerlines
- Turn the transmitter on (Controller)
- Turn on the receiver (Drone)
- Set the calibration of the compass
- Keep the drone level on the ground and proceed in a circle until the lights turn green (color may vary depending on the make and build of the drone)
- Spin the drone again until it turns green again
- The drone will blink colors and you are now calibrated

The above process will, of course, vary by drone manufacturer. If you are unsure how to do this, check the drone's manual or manufacturer's website for detailed instructions.

COMMERCIAL FLYING

In accordance with the FAA, commercial drone use is any use of a drone (quadcopter or otherwise) "in connection with a business". Essentially, if your drone is being used to provide a profit or a function for a business, it would fall under the commercialized drone category. Commercially used drones also have to be vetted through the Transportation Safety Administration regulations. However, as of today, no license or certification is needed for terra-firma or sea-based drones.

Commercial drones are used for a wide variety of projects, from surveys, aerial views for monitoring an area for change, security, photography, or videography. This identifies any type of business, whether large, small, or independent that utilizes the functions of a drone for business purposes.

- What to have on hand when flying
  - Part 107 certificate if you have taken and passed the exam
  - Driver's license
  - Insurance verification
  - Aircraft registration number
  - A summary of the Part 107 rules for reference
  - Flight operation manual for reference that explains:
    - Takeoff
    - Landing
    - Emergency Landing
- Communications sheet (phone numbers and frequencies of local authorities)
  - Maintenance log
  - Maintenance Kit
  - Knowledge of your aircraft's weight
  - Knowledge of nearby airports
  - LAANC authorization (only applies if you need LAANC authorization)
    - Application process
- Taking pictures and video
  - Auto: An auto setting allows for easily capturing of photographs and video. When using the auto feature, it allows the camera to be set to the optimal settings for your photo or video.
  - Manual: The manual setting allows the pilot to adjust the aperture and shutter manually for maximum control.

- Using headless mode
  Using headless mode is great for beginner pilots. It allows the beginner to control their drone easier in each direction while the drone orientation is tracked with its transmitter and not its general in-space orientation. This flying mode allows for a more effortless flying experience and is typically already available on many drones.
  - The activation of headless mode on many drones varies since many drones use different methods for activating the headless mode; in most cases, you will find the configuration details in your manual.
- Using FPV
  Many drones support FPV capabilities. FPV is acquired by putting on a pair of Virtual Reality (VR) goggles. The video feed from the drone is transmitted back to the goggles. The drone operator sees what the drone's camera sees. In most cases, the FPV operates on the 5.8 gigahertz frequency; the transmitter sends a signal to a receiver linked to an external monitoring system or goggles acting as the receiver.
  - FPV Goggles

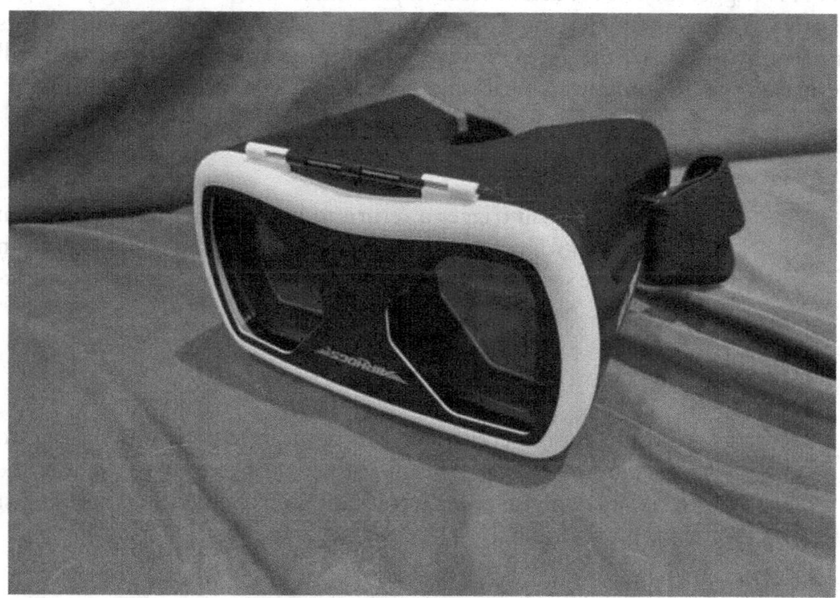

Image used with permission from Ralph M. DeFrangesco
DeFrangesco, R. (2021). First Person View (FPV) Goggles. Williamstown, NJ
First Person View (FPV) Goggles

FPV goggles could set the average user back anywhere from $50 to $500 or more. Makerfire has an entry level FPV goggle that operates at 5.8 gigahertz with dual antennas and a Liquid Crystal Display (LCD). The above FPV goggles get their input from a cellphone. The video is transmitted to a phone that is inserted into the goggles.

Orqa FPV.One Organic Light-Emitting Diode (OLED) FPV Goggles typically cost about $500. The FPV goggles have a 1280 × 960 OLED display with a 44° field of view. The Orqa FPV.One OLED FPV Goggles come with a head tilt alarm letting the user be aware that the goggles are out of optimal alignment. The unit also comes with a built-in defogger and support for many receivers.

FPV monitors are a cheaper alternative to FPV goggles. Operators can purchase a FPV monitor for as cheap as $40. There is even a FPV watch by Flysight that can be used as a monitor in a pinch. Of course, the watch only has a two-inch display but it operates at 5.8 gigahertz and has an LCD display.

- Return to Home
  - How to Return to Home Safely
    - How does your drone respond?

The return to home (RTH) is a feature that could save a drone when the RTH is triggered. The RTH feature is implemented when the Intelligent Flight Battery is triggered by low or erratic battery detection. When the battery hits a certain threshold, the RTH feature will be triggered, and the drone will return to the preconfigured location.

  - Failsafe RTH
    - Some drones will trigger a return to home if they lose connection with the controller for a given period of time. This could vary from three to five seconds. Some drones will immediately land if connection is lost. Not a good feature if you are over water or a crowd.
  - Smart RTH
    - A button on the controller can be used to return the drone to home for any reason. This might be because you lose sight of the drone, its flying in a dangerous area, is flying erratic, or you just want it back quickly.
- Obstacle Avoidance

Obstacle avoidance can detect obstacles in real time to avoid damaging collisions. The sensor that is a main contributor to the detection of objects when continuously scanning the open space for obstacles might not come preinstalled on your drone. This must be installed on your UAV and configured to work simultaneously with your controller.

- Enable obstacle avoidance under the "Active Track" section of the application on your controller.
- Enable RTH Obstacle Check to enable the feature that would allow your UAV to descend to a safe altitude while returning to home.

- Follow me feature

  - Some UAVs support a follow me feature. This allows a drone to follow a color or design while shooting video. Some pilots use this feature while skiing, snowboarding, or surfing. This is explained in detail in a future chapter.

- Stunts

A stunt drone is considered a quadcopter capable of having the ability to navigate in multiple directions and in many cases is done through a joystick or an additional featured button or buttons on your controller.

A stunt drone has the ability to complete a 360° flip or multiple flips. Other capabilities include pretracked acrobatic.

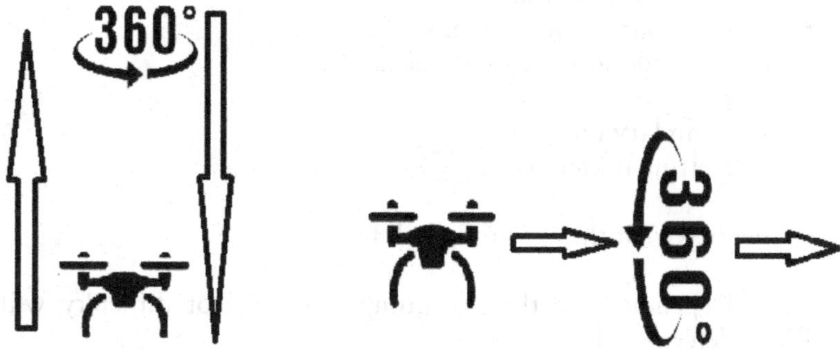

The above graphic shows which directions drones can spin in to perform stunts. Not all drones can do this easily, but most can be made to do some sort of aerial stunts.

- Auto takeoff/land

Many drones support auto takeoff and auto land. Auto takeoff and auto landing could be used in areas where the drone might land in water or a bad place if the drone got away from the pilot. The pilot registers where the auto land or home area is. If the pilot feels the drone is in danger, they could press the auto-land feature, and the drone will land where the drone was registered with the home position. Additional features can be enabled for landing a UAV safely, including landing protection and precision landing.

- Landing Protection can be enabled through an installed application or a manual controller. Landing protection allows a feature to be enabled to identify a suitable landing space for a touchdown.
- Precision Landings are enabled through a preinstalled application or a manual controller. Precision landing must be turned on and will attempt to land a UAV in the exact spot it took off when RTH was enabled.
- Landing/Takeoff
  - Instructions – Read the instruction manual that came with your drone or reference an online manual provided by the manufacturer.
  - Controller identification – what functions does your controller support?
  - Tracking – track your drone. It's easy to get disoriented when doing stunts.

- How to build an obstacle course
  - Determine the area size
  - Identify course timing (If any)
  - Identify indoor/outdoor course needs
    - Boundary lines
    - Netting if indoors
  - Identify what the age group will be
    - Depending on the age group, the level of difficulty will be determined
  - Identify a start and finish point
    - Creating a landing identification pad
  - Obstacles
    - User must move up/down
    - User must move side to side
    - User must be able to hover

  - User must be steady to maneuver through the obstacle course
  - User must remain in boundary line

- Point system

  - How are points earned
  - How are points deducted

How to build your own drone race course will be discussed in detail in the Do-It-Yourself (DIY) chapter.

CONTROLLER SPECIFICS

It would be impossible to discuss all of the controllers on the market and how they work. However, there is a benefit in discussing at least two types: a dedicated controller, one that comes with the drone when purchased, and drones controlled by a cellphone or tablet.

Most dedicated controllers are advertised as working at 2.4 gigahertz. Many people might recognize the 2.4 gigahertz range as the wireless access point in their homes and this is correct. However, dedicated controllers operate at 2.4 gigahertz, 5 gigahertz, or 900 megahertz for control and 1.3, 2.3, 3.4, or 5.8 gigahertz for video transmission. The lower the transmission rate, the further the signal can carry. Controllers that operate at 2.4 gigahertz typically have an operating range of 1 (1.6 kilometers) to 4 miles (6.4 kilometers), but much of this depends on having a full LoS.

DRONE INCIDENTS

As more and more drones take to the sky, there will inevitably be incidents. An incident can be anything from a crash causing severe damage to a drone flight over a cow pasture landing in a pile of manure. If it was reported, then most countries collect and use the data in statistics.

In February 2021, several drones were spotted over Frankfurt airport, canceling or delaying several flights. The polices spent several hours looking for the drones and their owner but could not find either. The police opened a criminal case and are looking for more eyewitnesses.

On February 10, 2021, Yemen's Houthi rebels crashed a bomb-laden drone into an Airbus A320 passenger jet at Abha International Airport in Saudi Arabia. The Houthi rebels are backed by the Iranian state and consider this retribution for the bombing of their country. The Saudi military does not know what type of explosive material was utilized in conjunction with the drone attack.

On January 23, 2021, a DJI Mavic Air 2 drone collided with a Chilean Navy helicopter. The two aircraft collided at high speed, causing the drone to fly through the windshield of the Bell UH-57B JetRanger III helicopter. One passenger in the helicopter was injured and treated for their injuries.

In December 2020, a Philadelphia, PA drone pilot was fined $182,000 (USD) for breaking multiple FAA regulations. The FAA was able to determine that the pilot broke at least 12 regulations over 26 flights. Legal experts guess that the FAA used YouTube videos that the pilot posted to determine the illegal flights and build a case against the pilot. The FAA warned the pilot multiple times about the illegal flights, but the pilot disregarded the warnings.

In October 2020, a drone interrupted a football match at New York Stadium in South Yorkshire, UK. The game was only 5 minutes into play when a referee spotted a drone flying over the field. Players were escorted off the field, and the game was halted for 10 minutes while the police was alerted. An investigation was launched, and the police was able to locate the drone and the pilot. The police confiscated the drone, and the investigation is ongoing.

In October 2020, a Georgia (USA) man was arrested for shooting at a drone owned by Georgia Transmission. Workers for the company stated that they notified the man that they would be flying their drone nearby, inspecting transmission lines. The man said he understood. The workers heard shots. When they landed the drone, they saw that the battery and part of the landing gear were destroyed. The workers called the police reporting the incident. The police contacted the man and he denied the incident. Police found a shotgun in his truck located on his property. The man was arrested and charged with reckless conduct and criminal damage to property. He will have to repair or replace the drone he damaged.

UA VISUAL PERCEPTION

Today, many drones come with Global Position Systems (GPS) technology built in. However, this feature is mainly available on high-end drones. Even if the drone has this capability, using it is another challenge in itself. Having GPS coordinate feedback means nothing without the capability to apply them against a map or vice-versa, using a map to program the flight path into a drone.

GPS capability only works when flying into a relatively open area or high enough that the operator will not fly into any obstacles. It is possible to fly a drone into a situation that the operator cannot fly out of using GPS. For instance, let us say you are flying into a heavily forested area. If you fly below the tree line, GPS will not be able to get you out. You would then have to take over manually and that would pose a challenge.

Keep in mind that when flying a drone and using a controller with a monitor, you view the world in 2D. Even if using FPV goggles, accurately measure the distance to an object is nearly impossible. Compounding this problem is that drones do not use 360° cameras, so there will always be a blind spot at any given time.

If, for instance, a drone is 500 feet away from the operator, it may become challenging to comprehend what situation the drone is in. Perception at that distance is inferior. Are you flying over someone or a car? Are you near a telephone pole or wire? How close is the drone actually to the building? Some drones do come with obstacle avoidance; however, that technology only goes so far. For example, the operator is flying the drone, and the obstacle avoidance system kicks in and auto lands the drone. You may think that is a good thing. However, since the drone is so far away, you didn't realize that it is about to land on a baby carriage or into a crowd of people.

One could argue that a person's quality of vision does play a part in this and it does. A younger person will see a drone further away better than an older person with poor vision. Also, the color of the drone, weather conditions, and drone speed all make a difference.

You will see your drone further away if you take the following steps:

1. Put bright colored stickers on your drone. If your drone is white, it will blend in and make it more difficult to see against a cloud.
2. Don't fly into fog, it will make it more difficult to see no matter what the color of your drone is.
3. Keep the drone on a sharper angle to you.

What this means is to fly higher versus lower. The sky is blue and white, so seeing a red drone will be easier than against objects on the ground.

Add lights to a drone, such as red or blue flashing lights work best and provide good contrast against the sky. Use a visual observer. Two sets of eyes are better than one. Operators can quickly lose a drone if you take your eyes off it to check your controller's battery life or indicators. Therefore, having a spotter or a co-pilot will help maintain consistent visualization of the drone, even if the operator has to glance down at the controller.

If you do lose sight of your drone, you have options. You can use the hover feature. Thus, allowing the operator to gain time to regain sight of the drone. The return-to-home feature will bring the drone back to the designated spot and allow it to regain sight. Finally, you can always use the autoland option. However, if you don't know where the drone is, auto land may cause more problems than just losing sight of it.

Although there are a few documented cases where a drone auto landed on a person, it does not mean that it did not happen; it is simply not documented. Most investigations involving drones list the incident as a crash, and this is because of the lack of experience of the people investigating the incident. When a drone's auto-land feature kicks in, you may not be able to take control of it. For instance, if you fly far enough away from the controller and lose the signal connected to the drone. On some drones, this will cause the auto-land feature to kick in. Additionally, there have been documented cases where pilots claim to have lost control of their drones once auto land starts its process. Some drones will return home when they lose signal with the controller.

MILITARY DRONE PILOTS

Recreational drone pilots are not the only ones to suffer from visual perception problems. Military drone pilots could suffer from spatial disorientation and vertigo. Researchers tested a group of military pilots flying drones from within a moving vehicle. The pilots did suffer from vertigo. However, drone pilots typically fly from within stationary trailers or buildings, which is less likely to happen. Spatial disorientation is another story. Pilots could become spatially disoriented for several reasons. Chief among them are control latency and lack of sensory indicators.

Latency is the time between when a pilot issues a command with the stick and the time it takes the Unmanned Aerial System (UAS) to respond to it. Remember, there could be thousands of miles between the pilot and the drone. There will be a natural latency between the two. Even under the worst of circumstances, a recreational or commercial drone pilot would most likely not experience these circumstances.

Researchers also noted a lack of sensory indicators. Flying a drone versus a piloted aircraft is different in that the pilot cannot hear how the aircraft is responding. Sound can indicate that there might be a problem, and they cannot feel the stick getting heavy. A drone pilot knows that there is latency and could miss a problem due to a heavy stick. Some pilots also noted that smell could be an indicator of a problem. A pilot inside a cockpit could smell electrical malfunctions or burning oil, whereas a drone pilot lacks that sense. A drone pilot only has camera vision and limited instrumentation. Vision is limited to where the camera is pointed, and we know you cannot cover all of the angles with a single camera.

Chapter 6

Hacking a drone

Image used from the public domain https://commons.wikimedia.org/wiki/File:RQ-170_Sentinel_impression_3-view.png
RQ-170 Sentinel impression

This chapter will discuss how to hack a drone. Hacking means different things to different people. Some people think of hackers as a teenager in a hoodie, eating junk food while sucking down gallons of caffeine-enriched drinks. I am sure some of those people exist. However, in the context of this book, a hacker is simply someone who modifies their drone to gain more capabilities by adding additional functionality from the baseline configuration, allowing the hacker to increase advanced, undocumented features.

Specifically, this chapter will teach you how to connect to a drone to download information, push data, and look at configuration files. As we have seen many times in the news, drone pilots have used drones to spy on people or invade the privacy of others. This chapter will teach how to identify a drone by sound signatures. After understanding this basic concept, you will learn how to control a drone using deauthentication techniques.

DOI: 10.1201/9781003201533-6

Finally, if you are a drone owner, you will learn how to protect your drone from being taken over by a nefarious hacker. There are steps you can take to protect yourself and your drone. The majority of work in this chapter centers around the Parrot AR and the Phantom 3, but the techniques can be used on other drones.

Many drones have the ability to Telnet into them. Telnetting into a drone gives you the ability to look at and change configuration files. This feature is also helpful for performing a forensic analysis, which we will cover later in this book.

THE TELNET PROTOCOL

Telnet is a protocol that runs on almost every computer platform today. It has been around since the early 1980s. Unfortunately, it is a very insecure protocol since it passes the credentials in clear text. This means if someone was on your network, they could easily sniff your login and password using readily available software.

By default, a Telnet server listens on port 23 for connections from a client. Your drone is the server, and your laptop would be considered the client. Most drones come with Telnet enabled. A much better alternative to Telnet is SSH. The SSH protocol allows a secure connection between your laptop and drone. If SSH is not enabled, you can easily enable it. People tend to use Telnet because it is easier to use, but you should use SSH where possible.

TELNET TO THE PARROT AR

In order to Telnet to the Parrot AR, first, establish a wireless connection to the drone. With the connected laptop, look for Ardrone2_256083 or something similar in the available wireless connection menu and connect to the device. It will not ask for a password.

The Parrot AR runs a version of BusyBox. You can use any piece of software that supports Telnet, such as PuTTY, MobaXterm, Solar-PuTTY, Xshell. ExtraPutty, etc. In this example, I will be using PuTTY. When you run PuTTY, you are presented with a configuration window. You can find out the default IP address that the Parrot AR uses by running a quick internet search. The search should return 192.168.1.1. In the Host Name (or IP address) box, type 192.168.1.1. In the Port box, type 23. This is the standard port for Telnet. Make sure the Telnet radio button is selected. Finally, click, Open.

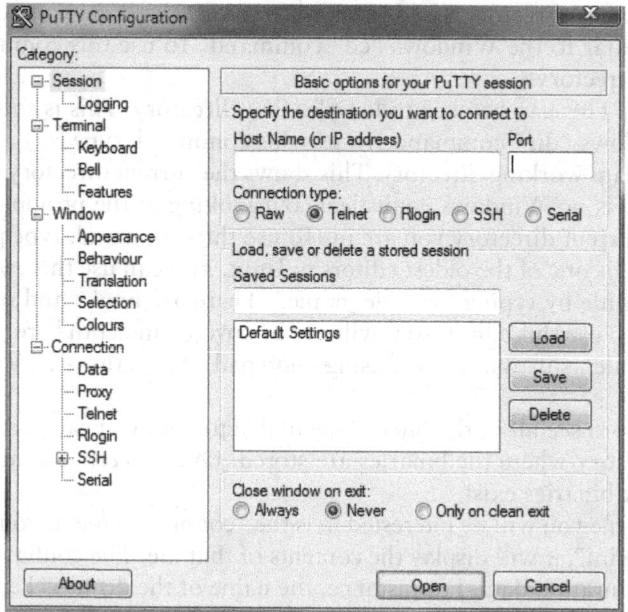

Image used with permission from Ralph M. DeFrangesco
DeFrangesco, R. (2021). PuTTY Dialog Box. Williamstown, NJ
PuTTY Dialog Box

You should see a pop-up box that states "BusyBox v1.14.0 ( ) built-in shell (ash)" or something similar depending on the version Parrot AR you own. When you connect, you are "root" on the drone, which means that you can do just about anything without restriction. As the saying goes, "with great power comes great responsibility". You can do damage to your drone by removing or changing things you should not be touching unless you know what you are doing. This book will not discuss administrative functions on your Parrot AR. However, this book will give a basic primer on how to traverse the file structure.

Start by typing "help" at the command prompt (#). Typing help will show the built-in commands the device uses. As mentioned earlier, the Parrot AR uses BusyBox, a Linux-like operating system, and uses very similar commands. The following commands might be helpful (all commands are in lower case):

cat – Short for concatenate. This command shows the contents of a text file. Windows uses the "type" command. To use this command type cat <file-name>.

cd – Change directory. Used to more from one directory to another. This is similar to the Windows “cd” command. To use this command, type cd <directory>.

ls – List. This can be used to list files in a directory. This is similar to the Windows “dir” command. To use this command, type ls.

pwd – Print working directory. This shows the current directory you are in. There is no Windows equivalent, but looking at the prompt will show the current directory you are in. To use this command, type pwd.

vi – This is one of the oldest editors in Unix. You can use this command to edit a file by typing “vi <file_name”. There are books and tutorials on how to use this editor, so it will not be covered in depth here. One of the equivalents in Windows is using “notepad <file_name>”.

If you want to see all of the binaries available to you, you can “cd/bin”. This is the directory where the binaries are stored. Once there, you can type “ls” to see what binaries exist.

Another file you will be interested in is the “config.ini” file. If you type “cat/data/config.ini”, it will display the contents of that file. The config.ini file contains configuration data. For instance, the name of the drone is kept here, the serial number, the date the drone was built, you can enable/disable the camera, set the time zone, display the name when connecting to WiFi, set the max/min altitude and much more. Your specific drones configuration file may vary.

TRANSFERRING FILES TO AND FROM THE PARROT AR

You can transfer files to and from the Parrot AR using an FTP client. Windows has a built in FTP client depending on the version, but you can use any FTP client you would like. In this example, I will be using the Windows FTP client.

In order to FTP to or from the Parrot AR, you need to establish a wireless connection to the drone. On the device you are connecting with, look for Ardrone2_256083 or something similar in the available wireless connection menu and connect to the device. It will not ask for a password.

You will need to open a command shell. Click the start icon in Windows. Type “cmd” at the prompt. These commands will vary depending on the version of Windows you are running. At the prompt, type “ftp 192.168.1.1”. The software will ask for a user name, just hit enter. You will be dropped onto the Parrot AR. Unfortunately, you have limited capability in FTP mode. You can list files using “ls”. By default, you will be in the/data/video directory. Thus, you can only get files that are in that directory.

To get files from the Parrot AR and put them on your computer, type “get <file_name>”. The FTP software will transfer the file to your computer. Whatever directory you type FTP at, is the directory your file will be in. Typing “quit” at the prompt will stop the FTP connection. Typing “dir” on your computer will show you the file you just transferred.

To transfer files to the Parrot, again make sure you are connected through WiFi. You will need to type "ftp 192.168.1.1" at the command prompt. The file you want to transfer to the drone must be in the directory you were just in when you typed "ftp 192.168.1.1". Once you are connected through FTP, type "put <file_name>". File_name is the name of the file on your system that you want to transfer to the Parrot AR. Any files you push to the Parrot will be put into the/data/video directory unless you move them to some other directory. Type "quit" to stop the FTP connection.

GETTING THE MAC ADDRESS

There might be times when you need the MAC to address versus the IP address. The MAC address is the hardware address, while the IP address is the logical address. In order to get the MAC address, connect to the Parrot AR using WiFi as described above. You will need a Telnet session. Connect to the Parrot AR using a Telnet client as described above. You can get the MAC address in a few places. The best place to get it is in the/data directory. Type "cd/data" once you have a Telnet session. The MAC address is in a file called "random_mac.txt". Type "cat random_mac.txt" and it will be displayed. The second place you can get it is in the config.ini file. If you are in the/data directory, type "cat config.ini" and scroll down to the MAC address heading.

TELNET TO THE PHANTOM 3

The Phantom 3 is similar to the Parrot AR in that you can connect to it using the Telnet and FTP protocols. This means you can view and change settings, download logs, pictures, and video right off the drone. You can do this using standard software tools or programmatically as part of something that you build.

The DJI Phantom 3 uses the following default IP addresses:

Controller – 192.168.1.1
Drone – 192.168.1.2
Camera – 192.168.1.3
Phone (DJI App) – IP is assigned. You will need to use a tool like NMAP to get this IP address.

The default credentials for the Phantom 3 are:

WiFi

Username: non needed
Password: 12341234

Telnet

Username – root
Password – Big~9China

By default, you cannot Telnet into the controller or the drone itself. You can enable these, but you can only Telnet into the camera initially. FTP is enabled on all interfaces, and you can log into each of these using the default credentials listed above. We will run through each of these to show how it can be done and what information can be collected from each interface.

Connect to your drone with your wireless network connection manager. If you are using a Windows operating system, you will need to use a Telnet client. There are numerous applications to choose from. For this example I will be using PuTTY. However, you can use:

Xshekk
Solar Putty
MobaXterm
mRemoteNG
SmarTTY
SuperPuTTY
ExtraPuTTY

Start PuTTY in a new window. Enter the IP address that you want to connect to. In this case, we will be connecting to the camera at 192.168.1.3. Set the port to 23, and make sure the connection type is Telnet. Click the Open button to connect to the device. You should now be connected to your drone.

Once connected, you can look at configuration files to check settings. Additionally, you can look at log data to get an idea of any problems you may be having and why. If you want to download the log data for further analysis, you can do this by using FTP which will be covered in the following sections.

HOW TO TURN ON TELNET

What if Telnet is disabled on your drone? The short answer is it is very easy to turn on telnet./etc/init.d/rcS is one of the most important network configuration scripts. It can load modules, configure the network, run applications and start graphical based interfaces. The script runs when the system is booted up. You will have to download the script and add a line to start up Telnet. To download the file, follow the directions below on how to use FTP. Once you have the file downloaded, add the following line to start Telnet:

```
telnetd –l/bin/ash &
```

This will start the Telnet service and move it in the background.

TRANSFERRING FILES TO AND FROM THE PHANTOM 3

Connect to your drone with your wireless network connection manager. Once connected, if you are using Windows, open a command line and type "ftp 192.168.1.3" and press enter. You should be connected directly to your Phantom 3's camera. You should now be able to download videos and pictures.

To download video and pictures, type "dir" at the command line. This will list all of the files in the directory. Type "get file_name" where file name is the name of the file you want to download to your computer.

If you are using linux, connect to the drone using your wireless connection manager. Open a terminal. At the prompt, type ftp 192.168.1.3. To get video and pictures, the process is exactly the same as with a windows computer type "get file_name".

In both the windows and linux examples, the video or the picture will be put in the directory that you ran the FTP command from. This is an important note since a lot of people forget where they ran the original command from. Of course, you could always look for the file, but keep in mind that a lot of data like this might not have a very memorable name. More than likely, it will be some long string of numbers.

OTHER DRONES

Not all drones support the Telnet or FTP protocols. You might have one last alternative to be able to connect to the device. Connect to your device using your wireless network connection manager. Open any browser. Try connecting to your device by putting in 192.168.1.1 in the url box. You can use any operating system for this.

If your drone supports the http/https protocol, you will see a screen that shows you are connected and will display data in your browser. At this point, you are limited as to what the drone manufacturer displays back to you. More than likely, you will not be able to change any settings in this view, but you might be able to view log data.

DEAUTHENTICATING THE PARROT AR

The deauthentication attack explained here was conducted on a Parrot AR, but the procedure should work on just about any drone that supports WiFi. The following procedures should be followed in specific sequence in order to deauthenticate the drone. This example was accomplished with a laptop running Linux.

1. Use your phone to see what Access Points are available. You should see the Parrot AR as one of them. Connect to the Parrot. There is no login or password.
2. Start the phone application for the Parrot AR. This can be downloaded from the Google store. It's called AR.FreeFlight.
3. Start flying the Parrot.
4. Using your Linux laptop with the Aircrack-ng suite installed, type the following commands:

    a. Airmon-ng start wlan0
    b. Airodump-ng mon0
    c. Airodump-ng –c 1 –bssid 00:00:00:00:00:00 –w drone mon0
    d. Aireplay-ng -0 2 –a bb:bb:bb:bb:bb:bb –c ff:ff:ff:ff:ff:ff –ignore-negative-one mon0

When you execute (d) from above, the drone pilot will lose control of the device. The application running on the cellphone will say, "no signal detected". If you do nothing, the pilot will gain control back in 5–7 seconds. However, if another person is running the same application and is connected to the Parrot AR, they can take over the drone when control is restored.

Four files are created when you run the above:

drone-01.cap
drone-01-csv
drone-01.kismet.csv
drone-01.kismet.netxml

It is not required to do anything with the files for this exercise. These are just supplied for reference purposes.

REPLAY ATTACK

It is possible to complete a replay attack on some drones. A replay attack allows an attacker to capture the pilot's signal to the drone and then replay it. A replay attack works best on cheaper drones. The reason is that more expensive drones use an encrypted connection between the drone and the controller.

You will need the following hardware and software:

Laptop (running Windows)
Antenna such as HackRF One, BladeRF, RTL-SDR, LimeSDR

Universal Radio Hacker

Use the following procedures (Windows):

1. Connect the antenna to the laptop and boot it up
2. Start Universal Radio Hacker
3. Select Spectrum Analyzer
4. Select the antenna from the device pulldown list
5. Select the frequency that your controller is transmitting on
6. Set the gain at 0, IF Gain at32and the baseband gain at 34
7. Leave the DC correction box checked and the Bias Tee unchecked
8. Click start

When you click on the controller, you should see a spike in the Radio Frequency (RF) readout in the top box and a spike in the waterfall box at the bottom. You will need to move the pointer at the top to adjust to the exact frequency your controller is transmitting at. Click on stop. Go back to the Main panel and click the record signal. The file will be pretty big, so only record a few seconds of the signal and save it out to a file using the save icon. If you change the default name, you will have to use the ".cs8" file extension.

The replay box will pop-up with the recorded signal in it. To replay the signal back, Change the modulation to "FSK", under additional parameters, check the additionally detect noise and additionally detect modulation boxes. On the send signal box, set the gain and IF gain to maximum. Turn off the controller and turn on the drone. Now, click on start. The drone should respond to the commands from the Universal Radio Hacker software.

Because of their size, low radar profile, and low noise, drones are hard to detect. There is a lack of a cost effective and automated way to detect drones. However, there is an increasing need to detect drones, especially when flown into restricted or unwanted airspace such as government installations, airports, and your backyard.

HOW TO IDENTIFY A DRONE

What if you needed to discover drones flying in your area? As an example, let's say you work for a government agency or a company and want to know when a drone comes near your facility? There are several ways to be able to accomplish this. The following sections explain just a few ways that will do just that.

Approach 1

Today, drones come with some very sophisticated equipment. You can purchase a Global Position System (GPS) systems, high resolution cameras and a myriad of sensors. A highly researched way to discover drones involves a solution that finds the direction of the control signal, and then disrupts the link to the drone. These types of solutions are costly and greatly depend on the geographic layout, and the amount of RF noise in the vicinity.

There have been many solutions proposed to address the problem of drone detection. Researchers have developed a detection system based on radar that can be expanded for the tracking, recognition, and imaging of drones.

One specific way is to use radio frequency wireless signals to detect drones. This can be further split into two solutions; one using active tracking that sends a radio signal and then listens for a returned signal and the second approach uses passive listening where it receives, extracts, and then analyzes a wireless signal.

The materials used to test this proposal are standard off-the-shelf hardware and software. The parabolic dish antenna was purchased off Amazon for under fifteen dollars. The software used to record the sound was Audacity and can be downloaded for free. However, any recording software can be used in lieu of Audacity.

A six-inch parabolic dish microphone, similar to the one below, was used as a primary collector.

Image used with permission from Ralph M. DeFrangesco

DeFrangesco, R. (2021). Parabolic Antenna. Williamstown, NJ

Parabolic Antenna

The parabolic antenna above has a microphone pointed in the center of the dish in order to collect signals. The antenna is powered by a 9-volt battery and has an earphone jack to listen to any output.

The antenna was placed facing upwards, towards the sky. When the test drone flew over the antenna, you could see the waveform increasing in intensity, reaching a peak, and then tailing off as seen in the following graphs. This indicates that a drone was approaching, flying near the collector, then continuing on.

The drone used for testing was a Parrot Mambo. Any drone could have been used for testing. This drone was used as a matter of convenience. The Mambo is shown below.

Image used with permission from Stephanie. DeFrangesco
DeFrangesco, R. (2021). Parrot Mambo. Williamstown, NJ
Parrot Mambo

The detector was mounted to pole thirty inches above the ground. This height was chosen for convenience. For the first test, the Parrot Mambo was flown over the detector at the height of two feet. The drone signature was easily identified as it flew towards the detector, as it flew over the detector, and then when it flew away from the detector. As we can see in the below image, the decibel level increased when the drone flew over the detector. A decibel is defined as a gain or loss in power. Decibels are typically used to describe values in microwave, satellite, and audio systems (Beasley, 2002).

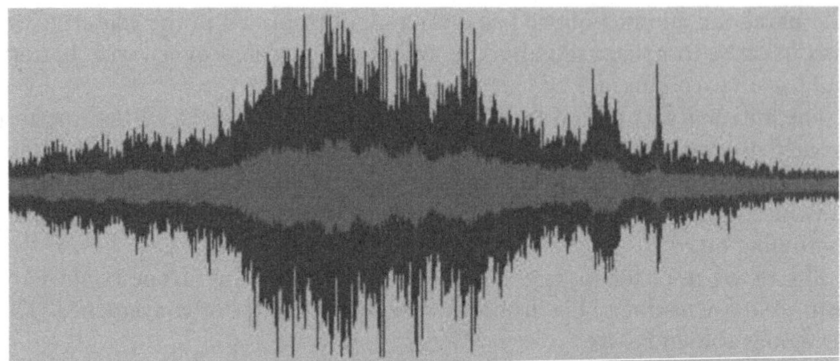

Image used with permission from Ralph M. DeFrangesco
DeFrangesco, R. (2021). Image of Measured decibels 1. Williamstown, NJ
2 feet above detector

In the second test, the Parrot Mambo was flown towards the detector at the height of six feet. The drone signature can be seen flying towards the detector, over the detector, and away from the detector.

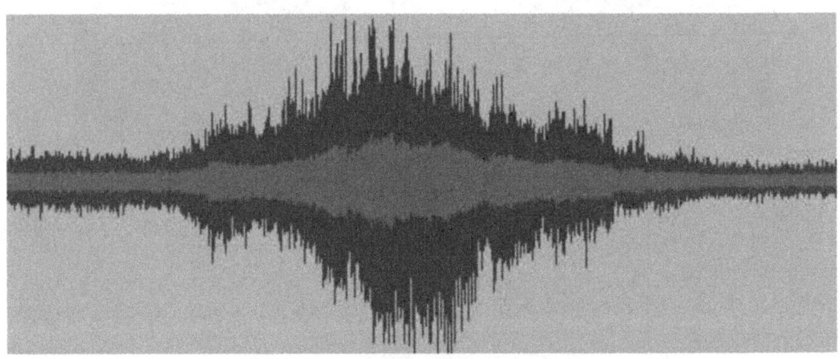

Image used with permission from Ralph M. DeFrangesco
DeFrangesco, R. (2021). Image of Measured decibels 2. Williamstown, NJ
6 feet above detector

In the third test, the drone was flown twenty-feet over the detector. See image below. The drone signature can be seen flying towards the detector, over the detector, and then away from the detector.

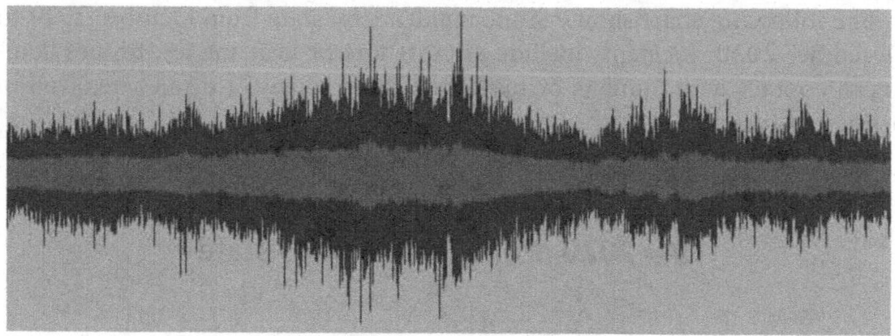

Image used with permission from Ralph M. DeFrangesco
DeFrangesco, R. (2021). Image of Measured decibels 3. Williamstown, NJ
20 feet above detector

For greater coverage many larger parabolic dish microphones could be used. These could be mounted on fences, buildings, and signs for the maximum coverage.

Since a drone uses a specific and identifiable acoustic signal, the system will not confuse a drone with a bird, balloon, or stray signal.

Approach 2

There has been an increase of drones in the skies over the last few years. The increase in drones is primarily due to drone manufacturers developing cheap products for consumer use. Availability and low cost make it difficult for the government to provide oversight over commercial operations. For less than $500 (USD), a consumer can purchase a drone that can take video at 2.7k HD and has a 12MP camera with a three-axis gimbal mount.

Consumers purchase drones for various reasons. Some people purchase a drone out of curiosity. They are not quite sure what the hype is about, so they purchase one to have it. Some people purchase them to do aerial tricks. Many drones are capable of doing air acrobatics. No matter what the reason, drones are a popular gift for many children and adults.

Along with an increase in drone, usage comes an increase in drone incidents. Drone operators have been behind many near collisions with aircraft in controlled air space. Drone growth is expected to rise significantly. The following chart shows worldwide drone sales by units sold through 2020.

Drone pilots frequently fly into no-fly zones. There have been numerous drone incidents where drones have flown into no-fly zones or near mid-air collisions. The question that law enforcement is trying to answer is how do we detect them?

The following chart shows drone incidents by state from October 2020 to December 2020. Incidents include aircraft hits or near misses, drones coming in contact with humans or buildings, and reports of drones involved in privacy incidents.

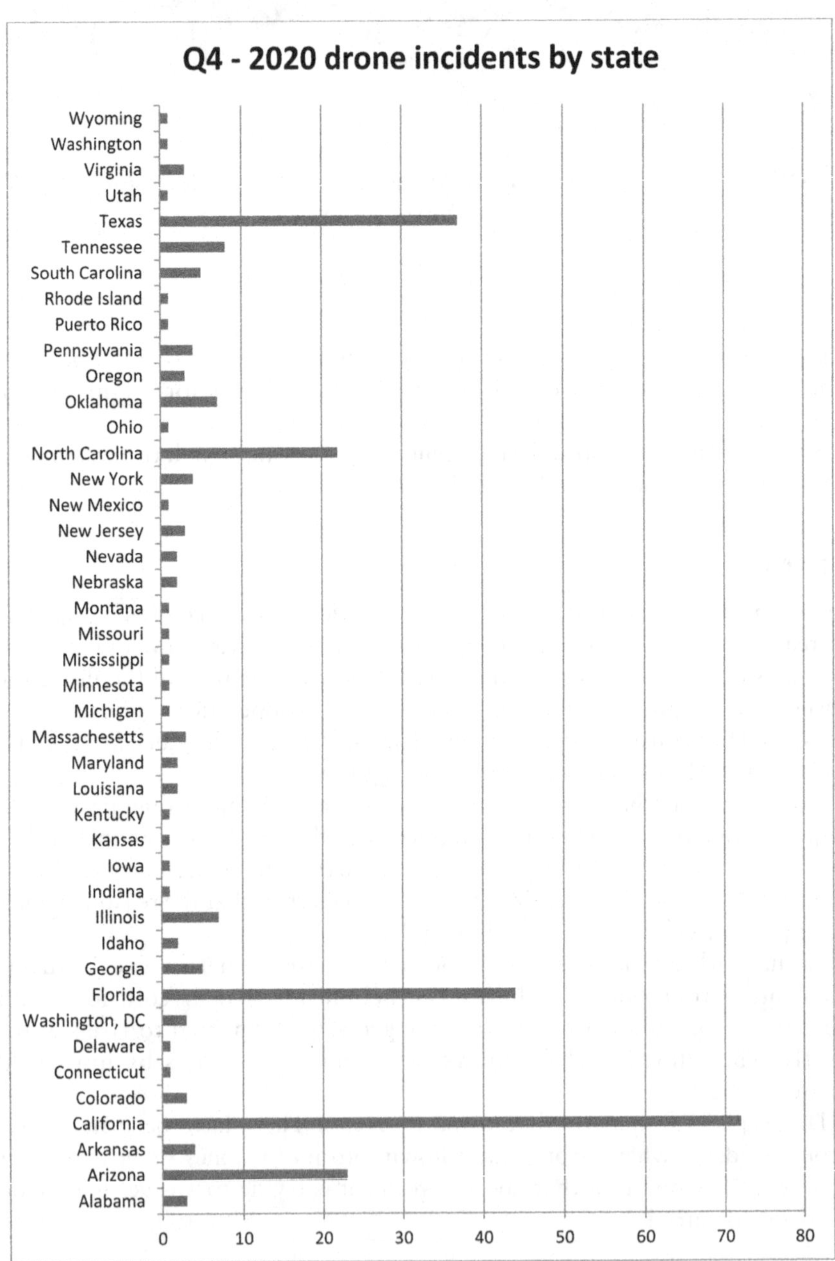

California has the highest incidents, followed by Florida, Texas, and Arizona. It is not surprising that California has one of the highest incident rates. First, California is a high-tech state that produces many drones. Second, it is a very popular state. This data was taken from the Federal Aviation Administration (FAA) website. You can get more data like this dating back to November 2014.

A drone, like any other machine, makes a sound. Sounds are not one frequency but are made up of multiple frequencies. It is possible to fingerprint a drone by capturing its sound and looking at the multiple frequencies contained in it.

The sounds from three devices were analyzed, a drone, a lawn mower and a helicopter. First, the. WAV files were down loaded. For this research, Zapsplat was used to download the recordings of existing sounds, since they were recorded with high quality. The recordings were played and captured in Friture using a screen capture of the output. Friture shows the dynamic frequencies of each file as it plays. The screen captures were put into MSPaint to identify the clear images.

The frequencies were then analyzed to see if they could be uniquely identified. Friture was used to analyze the frequencies because it shows the data in multiple formats, a scope, 2D, FFT and octave spectrums (Friture, 2018). For this paper, the octave spectrum option was used.

On the drone octave spectrum chart below, we can see that frequencies in the range 1–2 kilohertz and the 4–5 kilohertz are the most prevalent for this drone. This is expected because a drone has a high pitch to it.

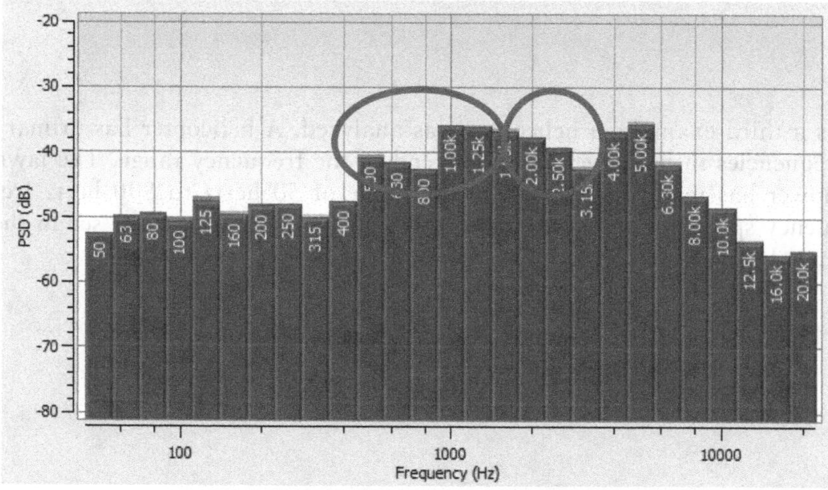

Image used with permission from Ralph M. DeFrangesco

DeFrangesco, R. (2021). Drone. Williamstown, NJ

On the other hand, a lawn mower should show more substantial frequencies in the lower and middle range of the spectrum. We can see that the frequencies that were more prevalent in the drone octave spectrum chart below are not in the lawn mower chart. The lawn mower has frequencies in the range of 400 hertz to 2 kilohertz are more prevalent for the lawn mower.

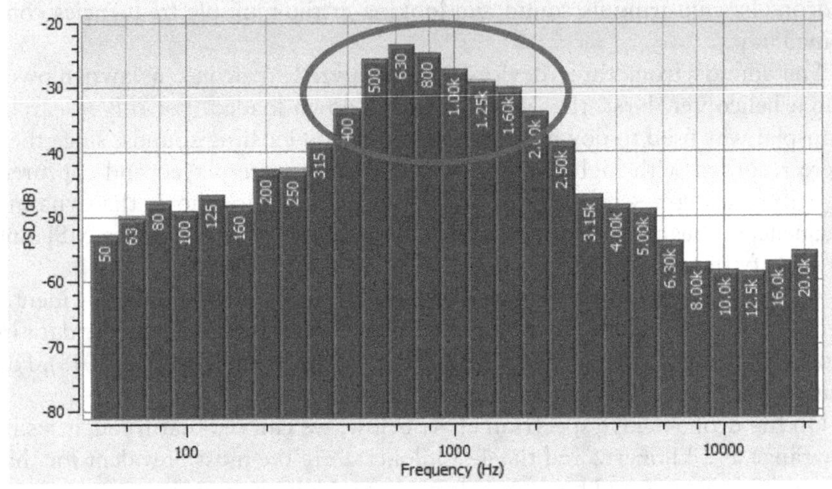

Image used with permission from Ralph M. DeFrangesco

DeFrangesco, R. (2021). Lawn Mower. Williamstown, NJ

Lawn mower

As a third example, a helicopter was analyzed. A helicopter has primary frequencies in the lower to middle end of the frequency range. The lawn mower has many frequencies in the range of 50 hertz to 800 hertz frequency spectrum and very few in the higher ranges, as we can see in the chart below.

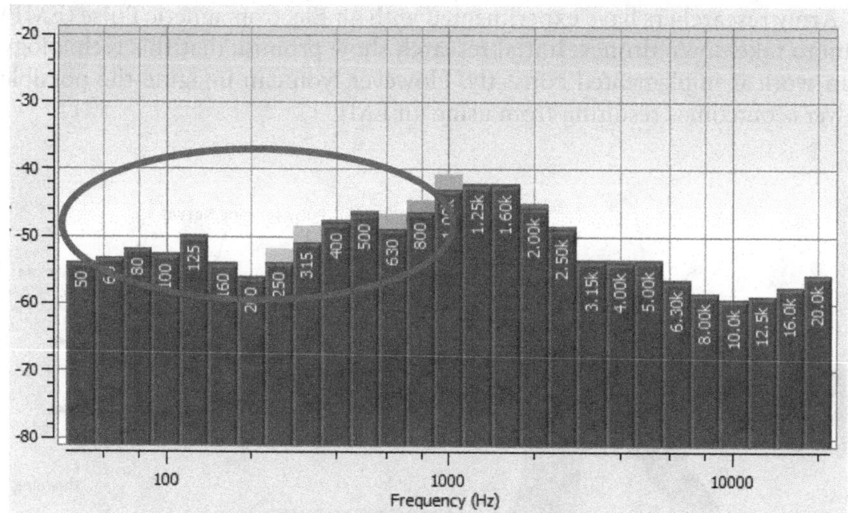

Image used with permission from Ralph M. DeFrangesco

DeFrangesco, R. (2021). Helicopter. Williamstown, NJ

Helicopter

We can see a difference in all three of the octave spectrum charts. Drones tend to have higher frequencies in their sound spectrum, while lawn mowers and helicopters do not.

Remember that not every drone, lawn mower or helicopter will create the same frequency fingerprint as shown above. However, through analysis, there should be frequencies that distinguish them from each other.

Drones have been involved in accidents where people have been injured. It is not out of the realm of possibility that a drone could kill someone, given the size and weight of some of the commercial drones available to consumers.

TAKING DOWN A DRONE

Drones have been involved in incidents many times with commercial aircraft. Legal authorities have tried numerous methods to try and take them down or detect who is flying them. Some methods authorities have used to take down a drone involve shooting it down with a special harpoon and net to “capture” the drone. This has had mixed results.

Army researchers have experimented with an Electromagnetic Pulse (EMP) gun to take down drones. Initial research show promise that this technology can work if implemented correctly. However, you can imagine the possible adverse outcomes resulting from using an EMP.

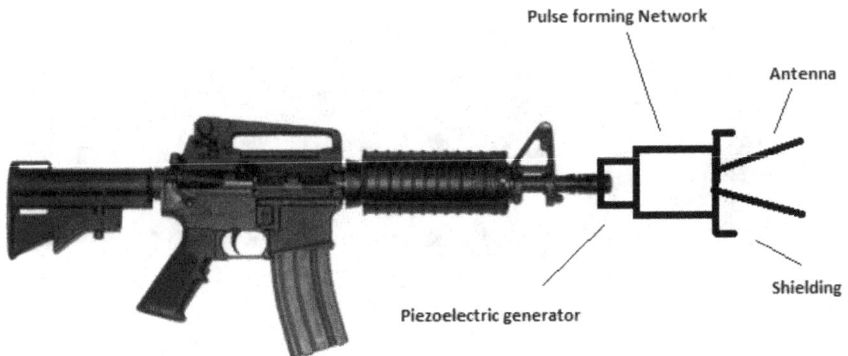

Image used with permission from Ralph M. DeFrangesco
DeFrangesco, R. (2021). EMP pulse gun. Williamstown, NJ
EMP pulse gun

The above graphic shows that the EMP pulse gun is based on an M4 rifle platform. The platform allows a ground-based soldier to fire a blank, which creates an electromagnetic pulse that is directed through the antenna to the desired target.

It is possible to make your own anti-drone "killer" gun. Several articles on the Internet explain how to make one out of a mosquito racket or a cheap throw-away camera. Caution should be taken as these are hazardous devices, and Federal law prohibits the operation, marketing, or sale of any type of jamming equipment. An EMP could take out your neighborhood's electronic devices, which you would be responsible for. Additionally, if someone near you were to have an electronic device in them, such as a pacemaker, you could disrupt the device and kill them. The following are some more examples that can be used to take down a drone:

- Bombarding it with signals
  - A drone can only respond to so many connections at once. If you overload it with too many connections, it is possible to cause it to shut down.

- Using a laser
    - Shooting a laser at the camera could destroy it
    - Cameras have sensors that change photons into electronic pulses which then become pictures. The light from a laser is intense and can damage these sensors. At the very least it would temporarily blind the camera.
    - This is difficult to accomplish on a moving target. You would have to fire it when the drone is hovering over a stationary target.

PROTECTING YOUR DRONE

We dedicated most of this chapter to detecting a drone, hacking a drone or how to take one down. However, not all readers will want to hack a drone nor do they want their drone to be hacked. In fact, they just want to fly their drone and be left alone. So, how can you protect your drone from someone hacking it? The simple answer is, do not take it out of the box, don't turn it on, and definitely do not fly it! Well, that's just not practical. Defensive security is the way to protect your investment.

Every drone has a computer in it. As we have seen all too often in the news, computers can be hacked into pretty quickly today. The remainder of this chapter is dedicated to protecting your investment.

Protect the controller

- Apply all up-to-date patches and updates
- Use anti-malware
- If possible, don't use it for other purposes
- Disconnect it from your computer when not updating it.
- If you are connected to a laptop, make sure your laptop is fully patched and updated.

Use a Virtual Private Network (VPN)

If you do need to connect to the internet, then use a VPN provider. A dedicated VPN connection will provide encryption between you and the internet. This lessens the likelihood that someone will attack your device.

Use strong passwords

Strong passwords will make it difficult for someone to take control of your drone controller. I recommend making your password 12 places long. This should be a combination of alphanumeric, numeric, and special characters. Weak passwords account for the number one way to hack a system

Use encryption

The best way to prevent someone from taking over your drone or getting to its data is to use encryption. Software like eMotion flight planning software uses Advanced Encryption Standard (AES) 256 encryption. For all practical purposes, this is not breakable by the average hacker.

Since this is a chapter dedicated to hacking, I have to mention that you can also write your own software to encrypt your communications link. Because of the speed needed, it would be best to write it in a language that uses a compiler versus an assembler. The "C" programming language is a great language to use for this type of application.

Buy a dedicated controller

A dedicated is best from a security perspective because it's harder to hack it. However, a lot of drones are sold that use WiFi. Using a controller that uses 2.4 or 5 gigahertz makes it harder to capture, analyze, and reuse a replay attack against your drone. 2.4 or 5 gigahertz are standard frequencies that most WiFi devices use. This means that your drone traffic will blend into every other WiFi device. That being said, someone could launch a Denial of Service (DoS) attack against that entire frequency and shut you down, along with everything else that uses that frequency.

Fly in a remote area

This recommendation is a double edge sword. Flying in a remote area means that there is no one around near enough to do any damage to your drone. This includes capturing your signal. However, if someone followed you or used a powerful enough antenna, they could more easily figure out your signal and can capture it because of the sparsity of signals in the area. The benefits outweigh the chance of someone capturing your signal.

Be aware of your surroundings

Good OPSEC (Operational Security), means being in tune with your surroundings. Do you see someone with a laptop or notebook near where you are flying? Do you see the same person around whenever you are flying your drone? Did a stranger just ask you for information about you or your drone? Call it paranoia, but people have been hacked using the exact scenarios described above.

SUMMARY

This has been one of the most interesting chapters to write. We are hackers at heart. So it was natural to extend hacking to drones. We covered how to connect to a drone and look at configuration files using Telnet. We also discussed how to pull from and push data to your drone using the File Transfer Protocol (FTP); two important capabilities to learn.

We learned how to deauthenticate a drone and take control over it. A replay attack allows you to capture the signal a pilot has sent their drone and replay the signal to take control of it. In order to take a drone down, you need to be able to identify a drone. We presented several ways how to be able to identify a drone using signal analysis. Now that you have identified it is indeed a drone, we discussed commercial tools to be able to take a drone down.

Finally, we discuss how to protect your drone from the above. We identified several ways to protect yourself and your drone from securing the controller, using a dedicated VPN, strong passwords, encryption, and being aware of your surroundings.

Chapter 7 discusses how to program your drone. There are times when you will want to fly your drone by a computer versus a controller. We discuss the many languages that can be used to program a drone and give several examples that include code.

Chapter 7

Programing a drone

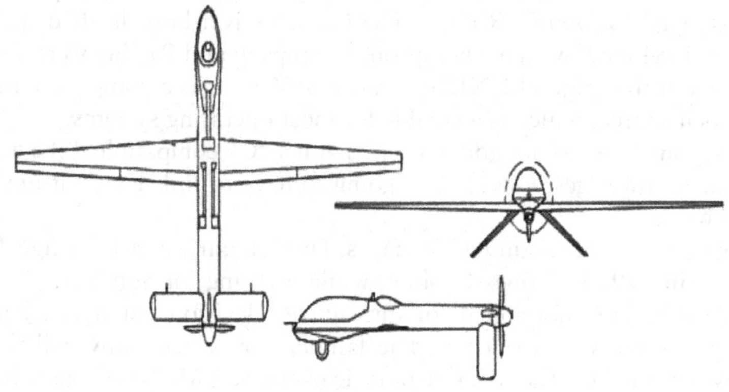

Image used from the public domain :https://upload.wikimedia.org/wikipedia/commons/c/c3/RQ-1_Predator_%28drawing%29.png
RQ-1 Predator UAV Drawing

This chapter will discuss how to program a drone, the programing languages available to program drones, some examples, including code how to program a drone, and some available software that brings out additional features of your drone.

Most drones can be programmed to fly specific flight paths. Usually, the cheaper drones will not have this feature. Drones like the Parrot AR and DJI Phantom make it extremely easy to program them because of the operating system and the support for the languages.

- What languages can you program a drone in?
  - Python
  - C
  - Java
  - NodeJS
  - Many more

DOI: 10.1201/9781003201533-7

Python is one of the most popular languages with programmers today. Python's history goes back to the early 1990s and was developed by Guido van Rossum. The language is an interpreted language that means no compiler is required and is open source available for most operating systems. The language is easy to pick up by most beginners and is a must learn language if your work in the Science Technology Engineering and Math (STEM) field.

Python can be used to program most drones. An example program is included later in the chapter to get you up and flying. Your drone's package is not going to say Python compatible. You will need to test to see if your drone can be programmed and identify what languages it will support.

C is an old language. Its humble beginnings go back to the early 1970s. It was developed by Dennis Ritchie. The language is a branch off of the "B" language developed by Ken Thompson. Thompson and Ritchie were also the developers of the original UNIX operating system. C is a compiled language and is open source, which is available for most operating systems.

Again, your drone is not going to say that it's "C" compatible. Like with all programing languages, you are just going to have to tinker with it until you get it to work.

In the history of programing languages, Java is a fairly new language. It was developed in 1995 by James Gosling, while working for Sun Microsystems. Java was originally developed for interactive television but it just had too much functionality at the time. The language was originally called Oak, which was named for the oak tree outside of the Sun Microsystems building. It was changed to Green, then to Java after a type of coffee from Indonesia.

Node.js is an open source run-time Java Scripting language that operates outside of a browser. Node.js is available for Windows, Linux, MacOS, and most Unix operating systems. Its libraries support programing a drone.

- Programing a specific flight path
  - Node.JS is a freely available application for download
  - A simple program to fly a drone programmatically, is shown below:

```
var drone = require("ar-drone");
var client = drone.createClient();

client.takeoff();
client.after(5000, function() {
       this.up(0.25);
       this.stop();
       })
client.after (5000, function() {
       this.clockwise(0.5);
       this.stop();
```

```
}). after(3000, function(){
       this.stop();
       this.land();
       });
```

CODE EXPLANATION

The above program is designed to do the following:

```
var drone = require("ar-drone");
```

Sets the variable drone type as "ar-drone". You will have to put your specific model/type in this field.

```
var client = drone.createClient();
```

Creates a variable client type. This makes the drone the client as opposed to the server.

```
client.takeoff();
```

This tells the drone to take-off.

```
client.after(5000, function() {
       this.clockwise(0.5);
```

This tells the drone to rotate 180° clockwise after 5 seconds.

```
}). after(3000, function() {
       this.stop();
```

After 3 seconds the drone to stop and stay in place after 3 seconds.

```
this.land();
```

Finally, this command tells the drone to land.

USING OTHER LANGUAGES

- Using a program to program the drone
  - In the following sample code, Visual Basic for Applications (VBA) is used to generate code to do the above. You could easily modify the code to develop a serious application that automatically generates Node.JS code by just clicking options.

- VB was chosen in this case since it is readily available on most desktops as part of the MS Office suite. Any language could be used to generate the same code.

Image used with permission from Ralph M. DeFrangesco

DeFrangesco, R. (2021). Code Generator. Williamstown, NJ

Code Generator

THE CODE BEHIND THE PROJECT

```
Private Sub CommandButton1_Click()
Dim myFile As String

myFile = "c:\Users\admin\Desktop\DroneApp.txt"

secs = TextBox1.Value

stopSecs = TextBox2.Value

degRot = TextBox3.Value


If CheckBox1.Value = True Then
       varDrone = ("var drone = require(""ar-drone"");")
       varClient = ("var client = drone.createClient();")

End If

If CheckBox2.Value = True Then
       CB2 = ("client.takeoff();")
End If

If CheckBox3.Value = True Then
       CB3 = "client.after(" + secs + "," + "function()" + " {"

End If

If OptionButton1.Value = True Then
       OPB1 = "this.clockwise" + "(" + degRot + ");"
       Else
         OPB1 = "this.counterclockwise" + "(" + degRot + ");"

End If

If OptionButton2.Value = True Then
       OPB2 = 1
Else
       OPB2 = 0
End If



If CheckBox4.Value = True Then
       CB4 = "this.land();"

End If
```

```
Open myFile For Output As #1

Print #1, varDrone
Print #1, varClient
Print #1, ""
Print #1, CB2
Print #1, ""
Print #1, CB3
Print #1, "this.up(.25);"
Print #1, "this.stop();"
Print #1, ""
Print #1, CB3
Print #1, "this.up(.25);"
Print #1, "this.stop();"
Print #1, ""
Print #1, "}) client.after(" + stopSecs + "," + "function()
{"
Print #1, OPB1
Print #1, "this.stop();"
Print #1, CB4
Print #1, "});"

Close #1

MsgBox ("Code Created")

End Sub
```

USING PYTHON TO PROGRAM A DRONE

In order to control a drone with a programing language, either the drone will have to support the language, or you will need to add an external board or two to the drone. The Raspberry Pi is the board of choice for controlling drones. If you add the Navio2 shield to the Raspberry Pi, you will have all the functionality you need to control your drone programmatically.

Image used with permission from Ralph M. DeFrangesco
DeFrangesco, R. (2021). Raspberry Pi 4 image. Williamstown, NJ
Raspberry Pi 4

The above picture is a Raspberry Pi 4 Model B. This specific model has 2 GB of Random Access Memory (RAM) and a Quad Core processor running at 1.5 gigahertz. It has multiple Universal Serial Bus (USB) ports; the two silver boxes on the bottom right, a gigabit Ethernet port, and the silver box on the top right, and have 802.11b/g/n/ac wireless capability. The long row of pins on the top is the Input/output (I/O) header. This allows you to control devices in the physical world.

Once we have the hardware set up, we will need to program our drone. The following is just a small snippet of code that will allow the drone to take off and then immediately land. This specific piece of code uses the DroneKit library. DroneKit makes developer tools for drones. The library can give add autonomous flight, live telemetry, and flight path planning through a Python Application Programming Interface (API) that support multiple platforms including Android, laptops, and web. According to the website, the DroneKit is currently free to subscribers.

- Add a little code in Python

```
From dronekit import connect, VehicleMode,
LocationGlobalRelative, APIException
import time
import socket
import exceptions
import math

def arm_and_takeoff(targetHeight):
      while vehicle.is.armable!=True:
      time.sleep(1)
vehicle.mode = VehicleMode("GUIDED")

while vehicle.mode!='GUIDED':
time.sleep(1)

vehicle.armed = True
while vehicle.armed==False:
      time.sleep(1)

vehicle.simple_takeoff(targetHeight)
      while True:
      if vehicle.location.global_relative_frame.
alt>=.95*targetHeight:
      time.sleep(1)

return None

vehicle = connect('127.0.0.1:14550', wait_ready=True)
arm_and_takeoff(5)

vehicle.mode=VehicleMode('LAND')
while vehicle.mode!= 'LAND':
      time.sleep(1)
```

OTHER WAYS TO PROGRAM A DRONE

- Follow-me feature
  - The follow-me feature allows drones to follow objects or colors. EZ-Robot makes free software that allows drones to take advantage of the follow-me capability. The concept is that you register the drone with a color or design. Once the drone sees the color or design, it will follow it.
- Using EZ-Builder
  - Download and install EZ-Builder from http://www.ez-robot.com/EZ-Builder

In the following example, we will use the AR Parrot. After EZ-Builder is installed, you will need to connect to your AR Parrot. Use the following instructions to connect to your AR Parrot:

1. Power on the drone by connecting the battery.
2. Using your PC or WiFi device, connect to the drone by selecting it from the wireless network menu.

How to connect to your drone with EZ-Builder:

1. From the start menu, select EZ-Builder
2. From the Project tab, select Add Controls
3. From the Third Party Robots tab, select AR Parrot Drone Movement Panel
4. From the Project tab, select Add Controls
5. From the Camera tab, select Camera Device

Now your screen should look like the following:

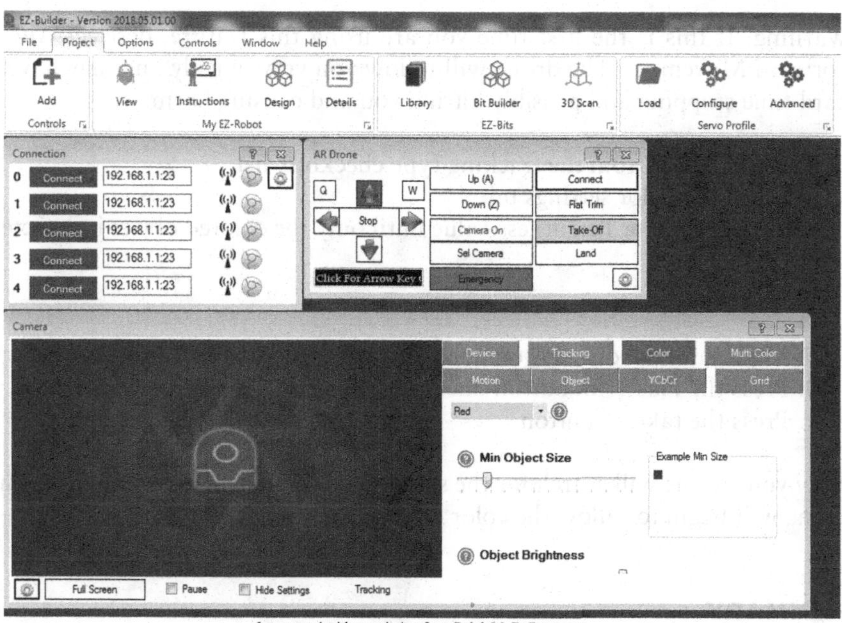

Image used with permission from Ralph M. DeFrangesco

DeFrangesco, R. (2021). EZ-Builder screen image. Williamstown, NJ

EZ-Builder screen

Let's connect to the drone using EZ-Builder:

1. Make sure that you are still connected to the AR Parrot with your WiFi device
2. On the AR Drone Movement Panel, press the Connect button
3. When EZ-Robot connects to your drone, the button will change to Disconnect
4. Press the Camera On button
5. In the Camera control, select, AR_Drone from the Video Device drop down

    a. You should now see the video image from your AR Parrot's camera.

Let's configure your drone to follow a color:

1. Make sure that you are still connected to the AR Parrot with your WiFi device and EX-Robot is still running
2. Press Config in the Camera control
3. Select the Movement Tracking checkbox
4. Select the Allow Left-Right Movement checkbox
5. Press Save

Warning: If this is the first time you are using this feature, deselect Allow Forward Movement. The drone will follow you very quickly and may have a hard time stopping and possibly hitting you and causing harm.

6. Select Color from the tracking type checkbox list.
7. Select the color settings tab.
8. Tune the Color Brightness value until only the colored object is detected

Let's take off

9. Make sure the drone is on a flat surface or a flat spot on the ground.
10. Press the Flat Trim button once
11. Press the takeoff button

Now your drone will search for the selected color or shape you taught it. The drone will begin to follow the color or design.

SUMMARY

In this chapter, we learned how to program a drone. We presented a few scenarios why you would want to do this. We discussed several languages that people use to program drones and even showed some code as examples.

We discussed how to use EZ-Builder to program your drone and how to unlock some of the features in your drone such as the follow-me feature.

In Chapter 8, we will discuss how to build your own drone. Building your own drone has gained popularity over the past few years due to Do-It-Yourself (DIY) kits and 3D printers; it's a very easy thing to do. We will discuss why you would want to build your own drone and give some examples. There are many DIY drone kits on the market and we mention a few of them and what they have to offer. If you decide to build your own, what tools will you need? We discuss an assortment of tools you should have on hand.

If you decide to build your own with a 3D printer, there are many materials to choose from. We discuss different materials and what they have to offer a drone hobbyist. After you print the body, you will need motors, propellers, a controller, and receiver. There are many on the market. We discuss what to look for when purchasing them.

Chapter 8

Build your own drone

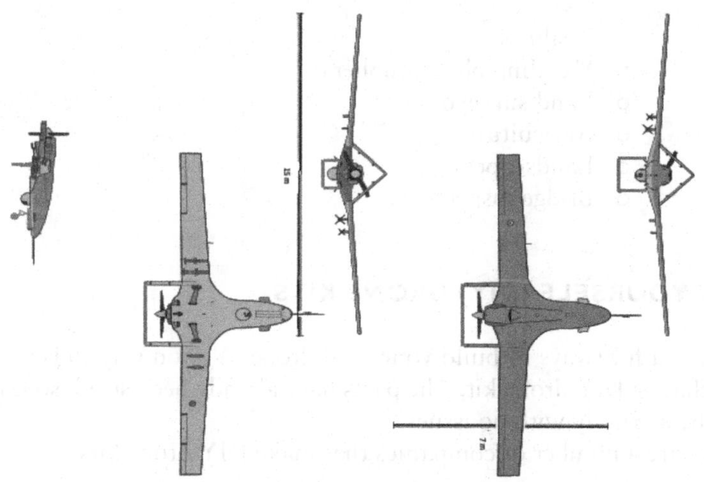

Image modified from the public domain: https://commons.wikimedia.org/wiki/File:Bayraktar-TB2-draw.svg
Modified Drone Draft

Today, building a drone is easy and fun. You might ask why you would want to build one versus purchasing one. There are a number of reasons to build your own drone as we have discussed in Chapters 1 and 5.

DRONE RACING

- Many people who race drones, custom build their own drones. They can get exactly what they want by doing it this way.
  - www.thedroneracingleague.com
  - Racing drones could start as cheap as $200
  - A top racing drone could run $1000 to $1500

DOI: 10.1201/9781003201533-8

- What makes a racing drone?
    - It's fast and maneuvers easily

VIDEOGRAPHY

- Many people that use drones to shoot video or pictures, build their own drones. This is because of the weight needed to carry cameras. Multi-rotor devices are used to handle this load along with custom sensors which can only be custom built.

    - Who uses a drone for photography?
        - Realtors
        - Wedding photographers
        - Land surveyors
        - Agriculture
        - Landscapers
        - Bridge inspectors

DO-IT-YOURSELF (DIY) DRONE KITS

There are a few ways to build your own drone. A good way to get started is to purchase a DIY drone kit. The parts have already been sized, so if it's built properly, it will fly with no issues.

There are a number of companies that make DIY drone kits:

GILOBABY – Makes a mini racing drone for beginners. The drone is easy to put together and costs around $30.

Hobbypower – This company makes a nice mini racing drone (H250). However, you will need a transmitter in order to fly this drone. This kit costs around $90.

QWinOut – Makes the Tarot hexacopter. This DIY kit is made out of carbon fiber. Although this kit is pricy at $650, it does include everything to get up and flying.

Most kits will not include the tools need to assemble the drone. It's a good habit to have a toolkit handy to make repairs to your drone. Some of the basic tools you will need are:

Exacto knife for trimming the parts

Small screw drivers. Jewelers screw drivers work nicely. Slotted and philips head

Needle-nose plyers
Magnifying glass
Finger-nail sanding board
Scissors
Paperclips

Spare parts such as propellers, batteries for the drone and controller, SD cards if your drone uses one for video, and picture storage.

There are a few things to keep in mind when purchasing a DIY kit. As indicated above, not all kits include a transmitter. Depending on what you want to do with your drone, you could purchase a cheap one just to fly around with, or a top of the line to compete with. The costs can vary from roughly $80 to $1000 or more.

Take your time when building your kit. A drone must be perfectly balanced in order to fly properly. Sometimes parts from a kit might have extra scrap on them from the molding process that might need to be trimmed off. Make sure that if you need to trim any wires, that you trim them all the same length where possible.

If you are building a quad-copter, you will be given four propellers. Two will be installed in the Clockwise (CW) direction and two in the counter clockwise (CCW) direction. You will need to pay strict attention to get these correct or your drone will not fly correctly.

3D PRINTING A DRONE

After building a few drone kits, you might want to custom build your own drone. Drone parts are readily available online or from hobby stores. People who build their own drones, frequently custom 3D print their own drone bodies to specific needs. The advantage is that it's relatively easy and cost effective to change body designs. Drone 3D body designs are free and are available by searching on the internet.

The PICO 110 High Performance Foldable Micro Quadcopter can be found on many sites that carry 3D printer designs. For this example, I downloaded the files from https://www.thingiverse.com/thing:2064676. When you download the zip file, there will be six files that make up the design; a top, bottom, and four connector arms. You will need to print two CW and two CCW arms. The other two files can be used if you decide you want First Person View (FPV). Since the body parts bolt together, you could always add this functionality later.

I used the DaVinci XYZ printer to print my parts but you can use just about any printer as long as the bed can handle the part size and the material you plan to use. The designer recommends using Polylactic Acid (PLA) to minimize warping, but the parts are so small, that the warping will be minimal. I used Acrylonitrile Butadiene Styrene (ABS) just because that's what was in the machine. With my setup, it took about 70 minutes to print all of the parts.

Once your model is printed, you will want to start to install the components to make it fly. The designer recommends using high performance Spintech Sidewinder 7 millimeter brushed motors. For the flight controller, there are a few recommendations including the Emachine FRF3 EVO with built in FRSKY receiver or the Emachine DSF3 EVO with built in DSM2 receiver. Whichever flight controller you choose, make sure it is compatible with your transmitter.

If you do not have a transmitter, the Spektrum DXe 6 channel transmitter or the FlySky FS i6 6 channel transmitters are both good choices. Hubsan makes good propeller blades. Remember to order two CW and two CCW blades. It would be beneficial to have extra handy just in case.

A variety of materials are available for 3D printing. Depending on the printer type, Nylon, ABS, PLA, metal infused, and a multitude of carbon reinforced materials are available to print with. A more complete list of materials is covered in the next section. Almost every part of the drone can be 3D printed with some exceptions. The motors, battery, and receiver/controller cannot be 3D printed at this time. This should not be a show stopper since the price of these parts is pretty cost effective to purchase.

Some other 3D printers that are good to use are:

The AnyCubic Mega is an entry level printer that uses a filament rack and will work with PLA, Thermoplastic Polyurethane (TPU), and ABS filaments. This printer costs roughly $279 US dollars. The bed is 8.27 × 8.27 × 8.07 inches.

R-QIDI technology makes the i-Mates printer. This printer is totally enclosed with an all metal frame. The printer supports PLA, ABS, TPU, and Polyethylene Terephthalate (PETG) and more. The i-Mates cost roughly $449 US dollars. The bed size is 10.24 × 7.87 × 7.87 inches.

QIDI Tech makes the model-X industrial grade printer. It has a large touch screen, WiFi, and can print a multitude of filaments including ABS, PLA, TPU, and flexible filament. The bed is 11.8 × 9.8 × 11.8 inches.

For those of you that want to build a 3D printer in order to build a drone, there are a number of options:

LABIST
Voxelab Aquila
Creality Ender
Geeetech
Tronxy

- Where can you get the drone designs?

- There are many sites that offer drone designs that you can download to print your own drone at home. These sites make. STL or stereolithography files available for download. Some top sites that offer. STL files include:
  - Cults
  - Thingiverse
  - YouMagine
  - Pinshape
  - MyMiniFactory
  - GrabCad
  - Autodesk 123D
  - 3Dagogo
  - 3DShook
  - Instructables

Designing your own drone is an undertaking in itself. However, many advanced drone pilots go this route because they see weaknesses in the drones they own and want to make them better. You could start by copying the drone you have. Take dimensions and build a design around what works. From there you can modify it. Be careful not to put that design on any public sites. Since the design is more than likely copyrighted, it would be a violation of their rights.

In order to design your own drone, you will need software to create the designs with. The following software is available for free download or online for free use:

- Design your own drone (CAD software)
  - Cura
  - matterControl
  - 3DPrinterOS
  - Slic3r
  - MeshLab
  - TinkerCAD
  - 3D Slash
  - 3D Builder

You could also use almost any Computer Aided Design (CAD) software for the design. However, you then have to convert the design to a. STL (stereolithography) file. If for example you are using AutoCad, it's pretty easy since AutoCad supports this file format natively. After creating your design, simply use the export feature and save it as a. STL file.

If you do not have AutoCad or CAD software that supports the. STL file format, don't worry. You can still convert it. There are numerous online sites that you can use to convert your design to a stereolithography format.

Once you have your design, you will now want to print it. What material should you use? There are many materials to build your drone with. Some materials are better than others to use, but it mainly depends on what your intention for the drone is.

- Body material options
  - ABS – It is highly durable, withstands higher temperatures but gives off hazardous fumes. It has to be used with a heated bed and shrinks a lot when cooling.
  - PLA – It is cheap, low melting temperature, does not require a heated bed, odorless, and is available in a myriad of colors.
  - Nylon – Is high strength, flexibility, and durability. Like PETG, Nylon absorbs moisture so it needs to be kept dry when cooling. It requires a high nozzle and heating bed temperature to extrude a part properly.
  - PP – Polypropylene is chemically resistant and food safe. An unfortunate feature of PP is that it is difficult to print and warps a lot.
  - PET/PETG – It (Glycol-modified) absorbs water easily from the air. This means that the part has to cool in a dry place to avoid shrinking. PETG is exceptionally good for parts that are subject to stress.
  - HIPS – High Impact Polystyrene is a hard material with elasticity of rubber. In 3D printing, it is typically used as a support material. That means that you can extrude another material with HIPS and then strip away the HIPS, exposing the structural material.
  - PVA – Polyvinyl Alcohol is soluble in water. Products that require them to dissolve in water, such as dishwasher pods can easily be created with PVA.
  - Metal – Metal infused filaments are available in a variety of metals including copper, bronze, iron, aluminum, and stainless steel.
  - TPE – Thermoplastic Elastomers have rubber like features. They are durable and flexible, which mean they can tolerate repeated stress and compression.
  - Wood – Filament manufacturers have come up with a way to infuse wood into PLA. Wood-PLA filament looks like read wood, but has the durability of plastic. Wood PLA comes in many simulated wood species including Pine, Birch, Cedar, Cherry, and Bamboo.
  - Conductive – Conductive filament can be used to projects requiring electrical conductivity. You can print circuit boards, sensors, and keyboards. Virtually anything requiring electrical conductivity and be printed with conductive filament.
  - Carbon fiber – Produces and extremely strong part that is very lightweight. One negative about the material is that it puts a lot of wear on the extruding nozzle.

- Motors

There are many companies that make motors for quadcopters:
AKK, EMAX, Readytosky, and Crazypony are just to name a few. A larger selection of motors will be recommended in Chapter 11, more on drones.

- Propellers
  - Drone propellers can be purchased in a variety of materials including carbon fiber, nylon, fiber glass, and titanium. Again, it really depends on your intent. Cheap plastic propellers are great for just learning how to fly. You will crash it many times, and plastic propellers take a beating. There is a bit of a double edged sword here because if you use carbon fiber blades, which are very expensive, and you crash, you just wasted your money. So choose the blades you use and what you plan on use them for wisely.

- Controller/receiver
  - Every drone needs both a controller, also known as a transmitter and receiver. Some transmitter manufacturers include Flysky, Hubsan, and iFlight. There will be more recommendations in Chapter 11, more on drones. Transmitters come in both 2.4 and 5 gigahertz versions and come in numerous channel varieties.
  - The receiver is an electronic device attached to the drone and uses a built-in antenna to receive the radio signals from the drone controller. The transmitter and receiver need to match. It's best to get the transmitter and receiver from the same manufacturer, although not mandatory. However, if you have a problem, then you only need to go to one place to resolve the issue.

BUILD YOUR OWN UNDERWATER DRONE

There are not a lot of underwater drones that the average consumer can afford to purchase. The ones that can be bought tend to be very expensive. Additionally, keep in mind that underwater drones do use a tether. So, when looking for an underwater drone, look for one that has a long tether.

It is possible to build your own underwater drone. I am in the process of building one as this book is being written (see picture below). The drone I am building will be made out of plastic tubing, some waterproof motors, a waterproof camera, and a tether that connects to a control box. The control box has a joystick that controls the drone's forward and backward direction and is connected directly to the motors and the power source. Diving and surfacing are controlled manually by moving the tether up and down.

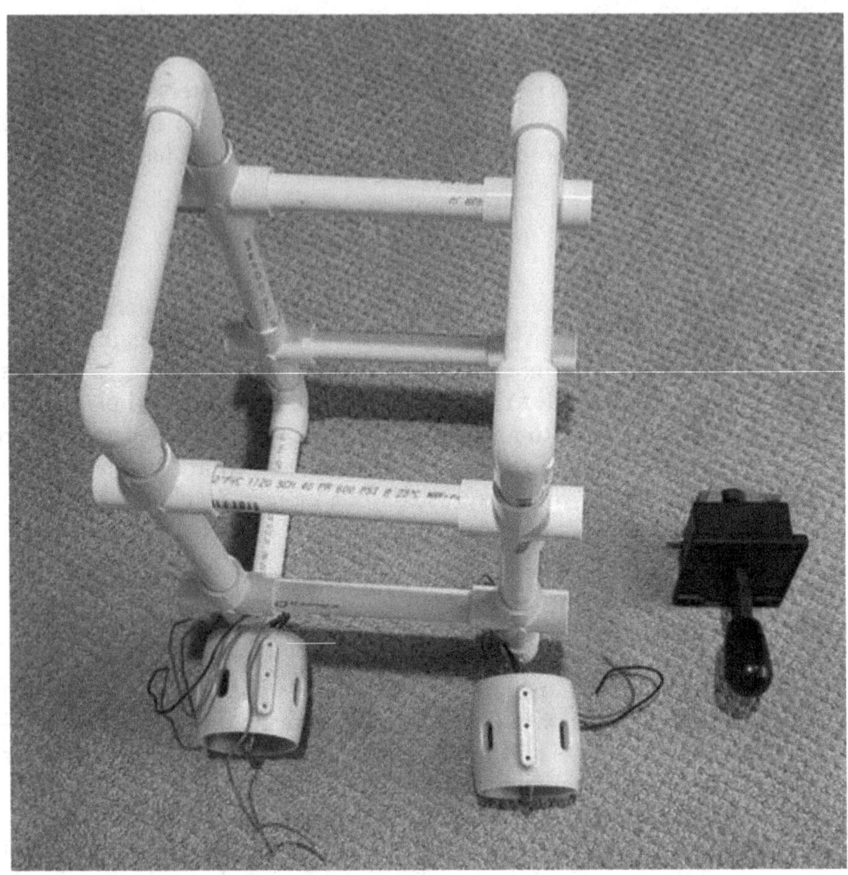

Image used with permission from Ralph M. DeFrangesco

DeFrangesco, R. (2021). Partially built underwater drone. Williamstown, NJ

Partially built underwater drone

The body is ½″ PVC tubing. The connectors are ½″ as well. After the drone is tested, you can weld the caps and connectors with PVC glue making it water tight. If the body floats, then sand can be added to the body to make is less buoyant.

The better underwater drones use a floating wireless access point. The idea here is that the drone is connected by a tether to the floating wireless access point. The drone pilot then connects wirelessly to the floating wireless point. Pictures, videos, and control data are transmitted through the wireless connection. This is a nice option to take pictures and videos without getting in the water.

Not all underwater drones use a floating wireless access point. Some just have a tether attached to a controller. This will allow people who want to snorkel or scuba dive take picture and video of their underwater experience. The only drawback is that you can't see the output until you download the pictures and video to a system for viewing. So if you thought you took a picture of you with a great-white, you will have to wait until the data is downloaded to see if you actually captured it or not.

A very cheap alternative

Some of you will still want to experiment with underwater drones but find the cost prohibitive and building one from scratch is just out of the question. If you just need to take pictures or video underwater, it's possible to do it cheaply. You simply need a glass jar, an underwater camera, silicone sealant, waterproof glue, and some strong string. I think you get where I am going with this. Put the camera in the jar in the on mode and taking video. Seal it with silicone sealant inside the lid as well as outside. Tie the string, then glue it to the lid and jar. You can now lower the camera into the water and it is taking video. Even if a little water should get into the jar, it's a waterproof camera, so it shouldn't be a problem.

SUMMARY

By using a DIY kit, anyone can build their own drone. For those of you that have a 3D printer, you can now easily print your own drone and purchase the parts easily online. We discussed why you would want to build your own drone and discussed where you can get a premade design to 3D print one yourself. We discussed the different materials that you can use to 3D print one and what parts you will need to complete it. Finally, we showed an underwater drone that we are currently working on and how you can build one yourself.

Whether you are an experienced drone pilot or just a casual weekend flyer, you can easily make your own drone.

In Chapter 9, we discuss DIY drone projects. Whether you purchased your drone, built one from a DIY kit, or 3D printed one, this chapter will show you different uses for your drone. There is a project that helps the seeing impaired, one to check on your kids, projects to build your own drone race course and many sensors and accessories you can add to your drone. If you like building your own drone, Chapter 9 takes it to the next level.

Chapter 9

Do-It-Yourself (DIY) drone projects

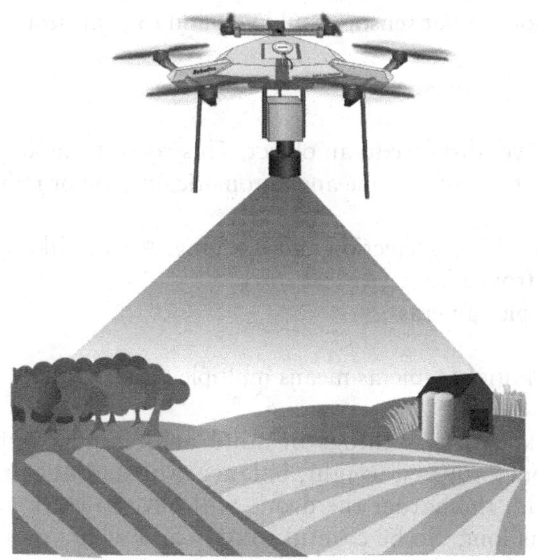

Image used from the public domain: https://commons.wikimedia.org/wiki/File:UAV_land_sensing_or_monitoring.svg
Landing Drone

So now you own a drone, but what do you do with it? This chapter will walk you through several projects that you can build yourself and also plants the seeds so you can take them to the next level. Some projects are easy to build and some you will find harder. The theme of this chapter is to use your drone for other things than just entertainment.

We will describe how to use your drone to help a sight-challenged person, how to use your drone to watch your kids when they are at the playground and how to build a drone race course. Some people are into prepping these days. Why not include a drone. We will discuss what you need to include in your drone bug-out bag. What if you discover a drone flying over your

DOI: 10.1201/9781003201533-9

property? What would you do? We discuss some alternatives. We also cover how to build a light show with your drone and how to use your drone to photograph you while skiing or snowboarding.

One big topic we discuss is how do you make money with drones? We cover jobs that drone pilots might be interested in as well as some websites that hire drone pilots.

WHAT ARE SOME OF ACCESSORIES YOU CAN MOUNT ON YOUR DRONE?

For the most part, drones are small. You will need a large enough platform to hold your sensors. What sensors could you add to your drone?

- Ultrasonic sensor
  - Gives distance to an object. This could be used to detect what's in front of a drone and automatically stop or redirect itself
- Lidar – Light Detection and Ranging. A radar like system that uses light from a laser.
- Multiple cameras
  - Multiple cameras means multiple photo angles

You need to keep a few things in mind before embarking on building your own drone projects; weight, balance, and aerodynamics. The more weight you add, the harder the drone will have to work and the battery will not last as long. So be careful to only add what you need to get the job done.

Keeping the drone balanced is very important. By default, your drone will work to try and keep things stable, but that only works to a point. Keep anything you mount on the drone as close to the center as possible. If you can't, then balance out the drone by adding something (weights, second sensor, etc.) to the other side. Also, keep in mind that you have to keep it in balance front to back and side to side.

You will quickly find that a stable, lightweight platform to mount your sensors on is very important. For many of the projects below, I used a diffuser found in many overhead lights. The panel is made of plastic and is lightweight because it has many square holes in it. The panel can be purchased at many home improvement centers for under $10 USD. I cut a piece of the panel to fit up under the drone between the four drone legs and used a zip tie to attach it to the drone.

Image used with permission from Ralph M. DeFrangesco
DeFrangesco, R. (2021). Drone Platform. Williamstown, NJ
Drone Platform

WHAT CAN I USE MY DRONE FOR?

Drones to help the blind

People needing visual assistance do not have too many options today. Companies offer the blind smart walking sticks. These devices can give obstacle feedback; detect pot holes, and changes in surface texture. Other manufacturers offer supposedly artificial vision and Global Position System (GPS) capability. Where they lack is detecting low hanging objects, sign boards, or open windows. Additionally, bumping into people is a common problem. There is no call for help if someone should need assistance except for the dependence on another person placing a call to law enforcement or Emergency Medical Services. This proposal will offer a solution that will do all of the above and give constant distance to object feedback, object identification, and make a call to a monitoring service or emergency help if needed. All of this will be accomplished with drones that fly and are terra-firma based.

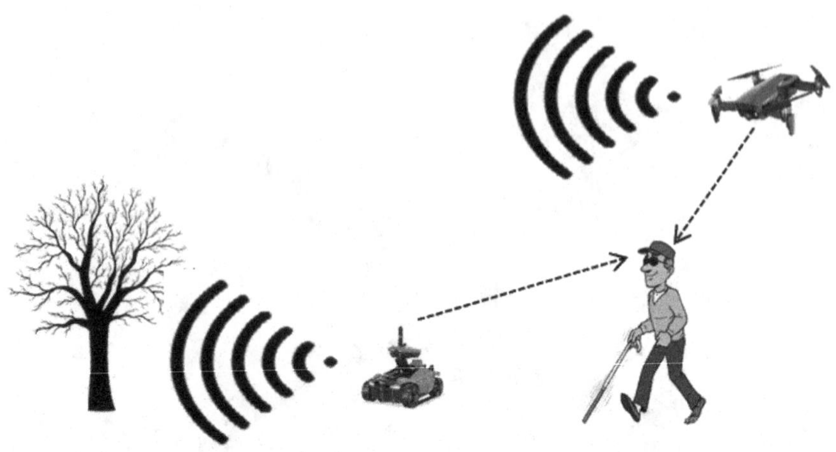

Image used with permission from Ralph M. DeFrangesco
DeFrangesco, R. (2021). Drone Potential. Williamstown, NJ

A man with a visual impairment is walking to the store. Aerial and terra-firma based drones follow him, giving feedback on the environment around him. In this scenario, the man would be alerted via a signal to his headphone that a low-hanging tree is in front of him. The distance and type of object is relayed so he can make course corrections. This is a fairly easy project to build and code. There are many resources to help you with this. Check with the following associations that are available to help the visually impaired:

- American Council of the Blind
- American Foundation for the Blind
- Blinded Veterans Association
- Foundation for Fighting Blindness

Drones to watch your kids

It's far too often that we see children abducted from play grounds. Another idea for the use of a drone is to use it as a monitoring system for your children. A drone can be programed to fly from your house to the playground your child is playing at. You could do reconnaissance on your child without them even knowing it. However, please keep in mind that you would be doing reconnaissance on other children as well. This might not be acceptable to the parents of those children.

Chapter 7 focused on programing a drone that you will need in order to develop this project. You can either use GPS coordinates or distance as long as you are not too far from the area you want to monitor. You will need a drone that you can download GPS coordinates to. Start by downloading map software for your area. Depending on the software you are working with, you can just click on the starting point, way points, and ending point, and download the data to your drone. When you start your drone, it will fly the coordinates you sent it.

A second option is to program it manually as described in Chapter 7. This is pretty easy to do. Start by roughly measuring the distance to the area you want to monitor. You can always adjust the distance by tweaking the value. If you don't want to use distance, you can program in GPS coordinates much like described above. Today, you can get GPS coordinate information from most cellphones.

Taken from the public domain: https://commons.wikimedia.org/wiki/File:Midorigaoka_Elementary_School_Mikicity_Athletic_ground.JPG
Local Park

The above picture is just a generic picture of an ordinary school playground. We can see some swings, a soccer goal, and climbing poles in the background. Regardless of the type of playground, the equipment or the location, a drone can be programed to fly there and capture video. There might be some challenges if the playground is covered with trees. In that case, you might have to be creative and fly closer. This might put you in jeopardy if you are flying over people so you need to be very careful how you fly over a playground.

Drones to watch your house

This project involves tying your drone into your outside camera system. If you have a whole house camera system, it might be possible to sync your drone into it. As an example, if camera one is watching your back door and it trips in the middle of the night, a program can be written to launch the drone to that specific spot to "put eyes" on whatever or whoever is there.

Here is how you get started. You will need to review how to program your drone as described in Chapter 7. So, you will need to program in the flight path around your house. You can follow the program listed in Chapter 7 and change the coordinates.

You will need a camera system that can act an alert. For instance, when your camera detects movement, it might turn on a light with a relay or perhaps it has a pull-up resistor. If there were one per channel, that would be ideal. You will need a raspberry PI and a network to connect everything together with.

If you only have one relay, then when the camera detects movement, it sends a signal to the raspberry PI's input (GPIO). The Raspberry PI then sends the signal to the drone to takeoff. The drone will fly the path around your house taking video or pictures the whole way. You will see what your drone sees and have it recorded to give law enforcement should you need to.

Personalized drone videos

Drones are used quite frequently in the filming industry for those hard to get shots in movies. Once, only the purview of helicopters and non-fearing pilots, drones became cheaper and better for movie makers. Drone pilots can fly drones into areas that helicopters cannot and don't complain about it.

You too can shoot your own videos with drones. You will need a drone, high-definition camera, gimbal, large SD card, good weather conditions, and a topic to shoot. Here is how to get started. Your drone is a flying camera, so you just need to point it in the direction that the action is in. If you are using a separate display for video, you will see what the camera sees.

When shooting video from a drone, slower movements are better. The movements should be smooth and deliberate. Remember you have an advantage over a regular camera, you can go vertical. So taking pictures from above adds a new dimension to your shots. Also, don't forget, you can use your drone to pan out and away from a scene, again difficult for a fixed camera to do.

Build a drone race course

It's highly unlikely that you will have access to a major league football or baseball stadium to create a real drone race course, but almost all communities have a little league or pee-wee baseball park. Optimally, you will need poles or markers that are least 8–10-feet tall. For a simple race course, put the markers 100 feet away from each other. You will need two people to make sure that each drone goes around the markers, and no one cuts them short. Put the drones on the ground, in-line with one of the markers. The pilots should stand mid-way between the markers. Use a whistle or air horn to start the race. You can race as many times around the poles as you want. If anyone cuts the markers short, then they get assessed a penalty.

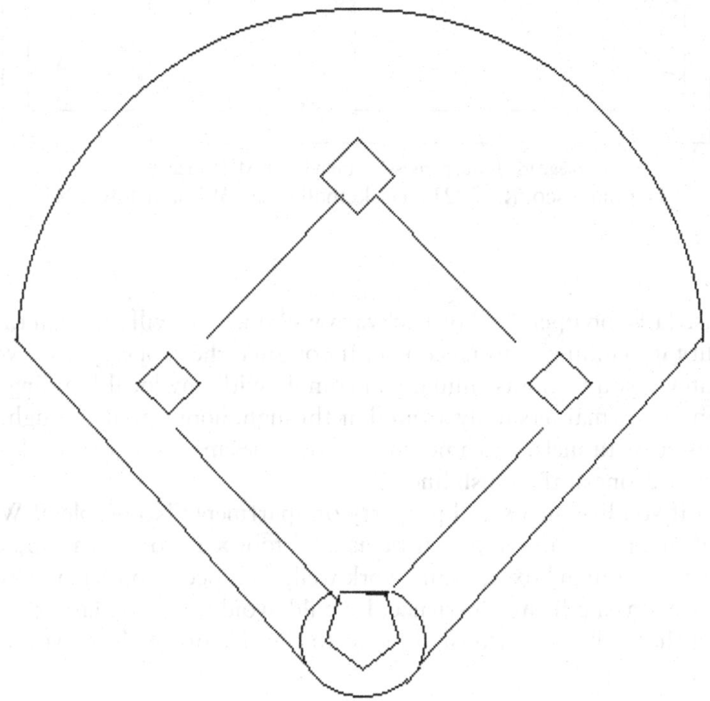

Image used with permission from Ralph M. DeFrangesco
DeFrangesco, R. (2021). Drone Track. Williamstown, NJ

An outdoor basketball court would work just as nice. The markers are built in and hard to cut short. Put the drones in the center and the pilot's should stand mid-way on the sideline. Again, use a whistle or air horn to start the race.

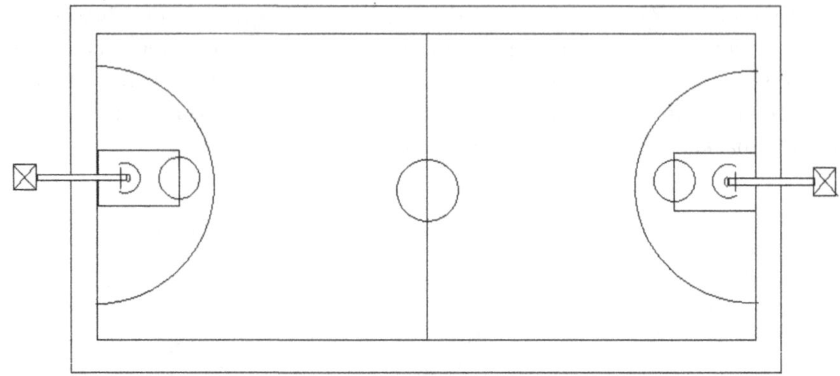

Image used with permission from Ralph M. DeFrangesco
DeFrangesco, R. (2021). Basketball court. Williamstown, NJ

If all else fails, on open field or backyard will do. This will give you the most flexibility in building a drone course. If you own the property, then you can do whatever you want, assuming you comply with any local building codes. You can put up markers to fly around or through, hoops, to fly though, boxes can be used as tunnels, and a net mound on poles makes a great back stop to catch your drone at the finish line.

What if you live on a small property or apartment? No problem! We have set up drone race courses in as little as a 10 foot × 10 foot area. You can fly around and through boxes, chairs work well, and soccer cones make for great markers. Speaking from experience, I would avoid using any lamps! I should mention that you are obviously limited to the size of the drone you can use indoors.

Drone bug-out bag

Are you a Prepper? Preppers believe that a catastrophe or massive disaster is likely to occur during their lifetime and they want to be prepared to protect their families and property. Whether you answered yes, no, or not sure you may want to consider including your drone in your bug-out bag. For those readers that are not Preppers, a bugout bag is a collection of supplies that you would take with you, should there be an emergency and you need to leave your primary residence.

The Prepper Journal, recommends using a drone as a force multiplier meaning that it will increase your efforts of being successful in a disaster scenario. A drone can be used for multiple purposes in a bug-out scenario. It's important to be able to see if your bug-out route is clear and passable. A drone can perform reconnaissance and show you exactly what is going on ahead of where you are travelling.

Drones can be used to deliver supplies when roads are closed. Although the capacity of a consumer drone is limited, commercial drones can carry a few pounds of food or supplies.

Perimeter security is extremely important during a disaster. Knowing if someone is coming at you is key to being prepared to defend against it.

If a drone does not have power, it can't fly. You will need to keep your drone charged and ready to go when needed. Carrying portable chargers is critical to keep your drone charged. You might need three to four chargers for this.

MULTIPLE DRONES

Did you ever want to fly multiple drones at once? Let's start with how to fly two drones at once. You will need more than one antenna on your laptop or notepad. Alpha Networks makes a nice antenna that's affordable and plugs in via a Universal Serial Bus (USB) port. It would be better to use Linux for this project, but it is possible to use Windows with some work. Once you plug in your second antenna, you will have two wireless connections available to you. Next connect each antenna to a separate drone wirelessly.

In Chapter 5, we learned how to program a drone. We can use the code we saw in that chapter to control both drones at once. The code is enclosed below:

```
var drone = require("ar-drone");
var client = drone.createClient();

client.takeoff();
client.after(5000, function() {
      this.up(0.25);
      this.stop();
      })
client.after (5000, function() {
      this.clockwise(0.5);
      this.stop();

}). after(3000, function(){
      this.stop();
      this.land();
      });
```

The definition in line one, "var drone = require("ar-drone");" will have to be changed for one of the drones if you are using two different drones.

Drone versus drone

What if a neighbor has a drone and they fly it over your property? This is a common issue that non-drone enthusiasts complain about. There are some irresponsible drone pilots out there that do not respect the privacy of other people. What if you could use your drone to take down another drone?

There are a couple of ways to do this ranging from the very primitive to the very sophisticated. As an example of a primitive method, you could hang a row of streamers from the bottom of your drone and fly in the direction of the drone, aiming to get the streamers caught in the blades of the invading drone. This would work best against an unskilled pilot, a non-agile drone or a drone that does not fly fast.

Remember that drones are nothing more than flying computers with wireless and video capabilities. What if you could capture the signal from the controller that is giving the drone commands and then replay them on your drone? It is possible to do this, but very difficult to pull off. If your drone is slower, then the best you can do is follow the other drone. You would need a faster drone that you wouldn't mind crashing into another drone.

Another option would be to use the follow-me feature that some drones have. You would have to have this preprogramed and ready to work. Using a picture of a drone could trigger the follow-me feature. Although you would never take the drone down, it could lead you to the person controlling the drone and then you could approach the pilot asking why they are flying over your property.

Drone + Pwnagotchi = a lot of cracked passwords

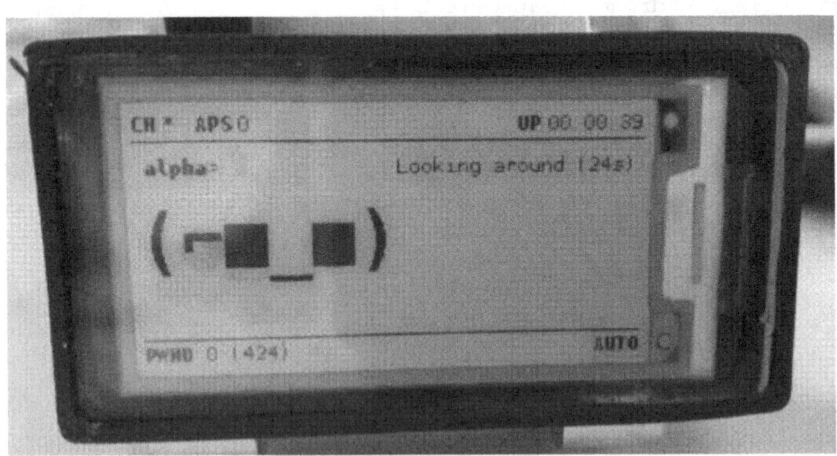

Image used with permission from Ralph M. DeFrangesco
DeFrangesco, R. (2021). Pwnagotchi. Williamstown, NJ
Pwnagotchi

A Pwnagotchi is a device that can be used to capture and crack WPA/WPA2 passwords. The Pwnagotchi is an open source project that is based on a Raspberry Pi Zero W. You will also need a screen (Waveshare 2.13 inch ePaper HAT), a large SD card, a battery pack, cables and a platform mounted to the drone to carry all of the above.

You will need to flash the Pwnagotchi.img file to your SD card. When you turn on your Pwnagotchi, you will need to keep it on for at least 30 minutes in order to train it. Once it's trained, you can mount the device on your drone, with the battery pack, and fly it around your neighborhood collecting passwords. Keep in mind that the passwords are encrypted, for now.

When you feel you have enough passwords, or the battery on your drone is about to die, fly your drone home. Take out the SD card and insert it into your computer. The Pwngotchi created Packet Capture files (PCAP) on your SD card. You can use Hashcat on Linux, MAC or Windows computer to crack the passwords. You must first convert the PCAP files to Hccapx format in order to use Hashcat. To crack the passwords follow these directions:

1. Get a copy of a password list like RockYou and put it into the Hashcat folder
2. Execute the following command:

    a. hashcat.exe –m 2500 PCAP.hccapx rockyou.txt

This could take anywhere from 30 minutes to 3 hours to crack depending on how much PCAP it has to process.

Drone fishing platform

Are you a fishing enthusiast? How about combining your love of fishing with your love of drones? This is a simple project to pull off. You will need the following equipment:

Drone (a good size one to lift a fish out of the water)
Drone pontoons
Fishing line
Bobber
Bait

Of course, you use any configuration that you want regarding the fishing portion. You will need to attach the fishing line across the landing rails. If you only attach it to one, you run the risk of tipping the drone over. Attaching it evenly to both rails will equal out the load. Leave several feet of line to the bobber. Land the drone on top of the water and wait. When you see the bobber dance up and down, you know you have caught something. Gently take the drone off and back to shore, hopefully with a fish onboard.

I would avoid flying the drone over the water while waiting for a bite. The noise will scare them off. Landing on the water with pontoons will provide a quieter environment. Also, avoid tying directly to the drone without a bobber. If a fish does bite and pulls your drone, it might pull the drone under water. That would be one for the fish and zero for you!

A drone light show

Many hobbyists use their drone for light shows. You will need the following hardware and software to be able to accomplish this:

Drone (large enough to carry the following devices)
Arduino UNO R3
1k Ohm resistor + red Light Emitting Diode (LED)
Breadboard with wires
9V battery with Arduino connector
Arduino Software IDE
Platform to mount all of the above on

Copy and paste the following Arduino sketch code into your Arduino sketchbook (IDE).

```
int led = 13;
void setup() {
  pinMode(led, OUTPUT);
}
void loop() {
  digitalWrite(led, HIGH);
  delay(1000);
  digitalWrite(led, LOW);
  delay(1000);

}
```

After you paste in the code, you will need to save it. Next compile the code. It should compile without any errors. Then simply upload your code.

This project can easily be expanded to include multiple color LEDs or an LED strip.

You will have to remember not to fly your drone at dark. The temptation is there, but it is illegal in most states to fly at night. However, you can fly up to dusk.

Take your drone snowboarding

What will you need?

- Drone
- High-definition camera
- Gimbal
- Follow-me software

There are a lot of examples of people using their drone to take footage of themselves doing sports. This is pretty easy to do. Chapter 7 explained how to set up the follow-me feature. Once you have the drone setup with this feature, launch the drone and it will follow you down the hill.

How to make money with a drone

Keep in mind that if you make money with your drone, you will need a pilot's license. That means taking and passing the FAA Part 107 test. The test costs $150 USD and if you pass, the license is good for 2 years.

Selling photos and videos

There is a large market for drone pictures. Remember, that not everyone has a drone. I used to live under a hot-air balloon flight path. I was very fastidious about keeping my lawn cut in a checker-board pattern. I would get many people knocking on my door asking if I know how good my property looked from the air. Someone always wanted to sell me a photograph of my property taken from the air.

If you want to make money with drones, you have to spend some money. The following are basics for drone photography:

Extra batteries
Filters for the camera
Photo editing software

Wedding photography

What better way to add to a wedding than showing pictures taken from a drone. With the wedding party outside, you can take-off and take pictures from all angles that a photographer could never get.

Selling drones

Reselling drones is a profitable business. Even though they can be purchased online today, many people want to see the drone and fly it before buying one. You as a drone reseller can offer this to your clients.

Drone clubs – Join a drone club. There are many drone enthusiasts out there and they join clubs to be with other drone enthusiasts.
Drone events – Setting up a table at a drone flying event is a great way to generate drone sales.
Drone conferences – You may not sell many drones at a conference, but it is a great way to get your name out there.
Online sales – Evidently, people look to the internet to purchase just about anything. Having an online presence allows for additional publicity and sales that can be placed around the world.

Working with a realtor

Drones offer a new dimension to viewing a house. With a drone you can fly the property, showing the potential buyers what the property looks like.

Working as a drone pilot

There is not a big need for professional drone pilots at this point. This will change in the future as more companies adopt drones for commercial use. You will need your Part 107 license, there is no getting around that.

Sites that offer employment to drone pilots:

www.soldbyair.com
https://dronebase.com/pilots
https://www.droneseed.com/
https://www.eversource.com/

If you are still interested in becoming a drone pilot, then the following industries use drones and are good sources to look for employment:

Agriculture
Construction (Inspection)
Bridge Inspection
Forestry
Mining
Gas (pipeline inspection)

SUMMARY

We covered how to create a home-made drone race course, sensors you can add onto your drone, taking your drone fishing and skiing, and how to make money with your drone. How to make money with drones has been a huge topic of interest. The job market is small, the interest and talent are big. There are ways to make money flying drones, you just have to spend the time looking for it and have patience. One option is to start your own business rather than fly for someone else. Whether you fly for someone else or start your own business, you will need a commercial drone pilots license which you can get from numerous tech and secondary institutions today.

Chapter 10 is a very specialized topic. Most people that fly drones don't even think about how forensics applies to what they are doing, but it does. Unfortunately, we only hear about it when someone does something they shouldn't with their drone like flying near an aircraft or crashing into a crowd. This chapter is meant to help those who perform forensic investigations or someone interested in working as a forensic analyst. We have included a template showing the questions to ask and the evidence that should be collected after a drone incident.

Chapter 10

Drone forensics

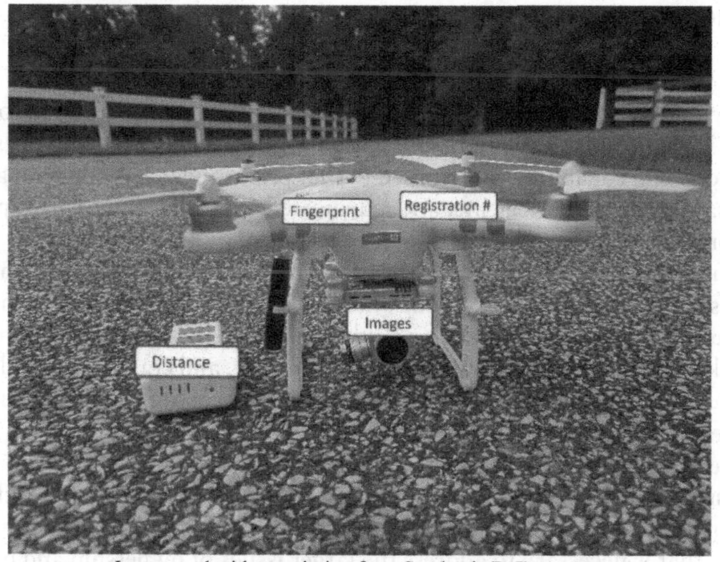

Image used with permission from Stephanie DeFrangesco
DeFrangesco, R. (2021). Forensic Image. Williamstown, NJ
Forensic Image

This chapter is dedicated to drone forensics. You might be wondering why you would ever want to perform forensics on a drone. Unfortunately, we have seen drones in the news that have caused problems at airports, privacy concerns, and have even crashed and caused physical harm to people and property.

When we think of forensics, we think of a Medical Examiner trying to figure out how or why a person died, or someone trying to pull fingerprints off of a door or piece of furniture. The basic premise of forensics is based on science and facts.

DOI: 10.1201/9781003201533-10

When a drone crashes into a crowd and injures someone, if the owner of the drone is not apparent, then we will have to find who indeed owns it and why it crashed. It could have been an accident, intentional, or a malfunction of the drone. Additionally, someone must be held accountable for causing the injury.

Performing forensics is not easy. It requires years of education and experience. We have tried to give you some criteria to ask questions to get you started. The checklist discussed later in this chapter is a great starting point. We can direct you as to what questions to ask, but in the end you will have to do the analysis.

INTRODUCTION TO FORENSICS

There has been a huge proliferation of drones over the last few years. This is mostly due to a reduction of hardware cost and ease of availability. Today, a drone for consumer use can be purchased for as little as $20. Of course, the functions and capabilities are limited in that price range. A drone with an HD camera, GPS, long battery life, and a dedicated controller can easily top $800 or more. Cost is directly related to options added and capabilities.

Consumers purchase drones for various reasons. Some consumers are not quite sure what the hype is about and so they purchase one just to have it. Whatever the reason, drones are quickly filling the skies. Many consumers do not take the time to learn how to fly them properly, and crash into and destroy private property or fly into restricted air space. There are no established procedures for conducting a forensic review of a drone.

Forensic researchers have not developed a specialized approach to drone forensics, which may have future legal implications. This research will develop a set of forensic procedures, which can be applied to a broad range of drones that is necessary to address the larger issues of ownership and mission purpose. This work is paramount in the fact that drone sales are rising rapidly and are expected to reach over 3 billion by 2024. More drones in the sky will equate to more incidents and the need to identify issues like ownership, mission purpose, and responsibility.

Research on this topic will focus on the forensic techniques, exploits, and data collection analysis of drones in the DoD Group 1 category, small drones while producing a set of standards for a forensic investigation. This study can help identify whether the procedures developed for forensic investigators can be applied to other categories of drones. As an example, forensic techniques already exist for performing analysis of removable media. Drone forensics offers a unique opportunity, which requires unique forensic techniques to collect and analyze information about the whole drone.

Not all drones have the same options or operate the same way. Some drones are controlled by Wi-Fi devices and cellphones. Some drones have a dedicated controller, which operates in the WiFi (2.4 gigahertz) range, but are not considered true WiFi devices. Some have on-board storage while others do not. Some drones have removable media and other devices you must connect to the device to download the data. Many drones have some kind of camera on-board and some come with multiple cameras and top-of-the-line optic packages to enhance image capture.

A loss of opportunity analysis would quickly show traditional forensic media analysis will only go so far in investigating drone incidents. Drones are quickly filling our air space so a deeper, more focused, long-term solution to drone forensic analysis is needed to analyze the opportunities that drones have to offer. Without a forensic analysis technique specifically geared towards drones, forensic investigators will undoubtedly be trying to fill a large void in analysis possibilities, which no other forensic technique can fill. As an example, the direction the drone came from could be valuable information. Digital media analysts are not trained to look at this information. However, it could play an extremely important role in determining whom it belongs to and what its mission is.

Key questions to answer are:

1. How do you best determine drone ownership?
   It is very difficult to determine who owns a drone due to the fact that although it is mandatory to register a drone, there is nothing in place to force registration. Additionally, a drone can be flown from miles away making it more difficult to determine who flew it. Some drones can be programmed to run a specific flight path without an operator intervention at all.
2. What information can be collected to determine drone ownership?
   Drone operators flying a drone weighing >0.55 lbs but <55 lbs, must register their drone with the FAA. Failure to register can result in a fine and/or criminal charges. This registration information can be used in a forensic investigation to determine drone ownership. However, outside of a monetary penalty and a possible criminal action; nothing forces a drone owner to register it. The last thing on a criminal or terrorists mind is to register their drone with the FAA.
3. What was the main mission of the drone?
   A hacker can disable security features on a drone to do things such as fly the drone in a no fly zone or allow the drone to fly higher than it was intended. A forensic analyst may never know the exact mission of a drone; however, they are left with the aftermath, trying to assemble the pieces back together to determine what happened.
4. What private information was collected by the drone?
   A major concern regarding drone usage is they could record images of people. Drones could be outfitted with sensors such as a high-zoom lens, microphone, and a Global Position System (GPS), which would allow additional personal information to be collected on people. Lawyers could argue this might be a violation of privacy.

Evidence collection, analysis, and correlation are key to determining who owns a drone and what its mission was. When there is an incident involving a drone, forensic analysts are the last person thought of. It's not the fault of law enforcement. They are often forced to deal with new technologies without having the proper training or tools to deal with these situations.

A Forensic Analyst with drone experience can be invaluable when there is an incident involving a drone. They will know what evidence to collect and how to collect it. It's important to treat the accident scene like any other crime scene. That means not touching anything, especially the drone. The crime scene still needs to be processed, but the drone and associated devices need to be processed by someone with experience processing digital media and avionic data. Most local law enforcement agencies do not have this experience.

The following evidence collection form can be used to collect the data that is in need of processing. Keep in mind that this form is only for evidence collection and not analysis. A trained analyst will be able to look at all the evidence as a whole and answer the questions who owned the drone and what its true mission was.

Evidence collection form

Direct evidence	*Collection examples*
Real or physical evidence	Laptop
	Cell phone
	Tablet
	iPad
	Drone controller
Demonstrative evidence	Maps of crime scene
	Mapping software
	Photographs of crime scene
	Video of crime scene
	Diagrams of crime scene
Documentary evidence	Drone manual
	Assembly directions
	User manual
	Pics/Video
	Custom application code
Questioned documents	Journal entries
	Threating letters
	Confessions
	Computer links
	Witness testimony

(Continued)

Circumstantial or indirect evidence	
Scientific evidence	Fingerprint identification
	Drone serial number
	Drone manufacturer
	Drone make/model
	Sales receipt
	Attached sensors
	Accessories
Empirical evidence	Temperature
	Weather
	Location
	Date/time
	Direction drone was found facing
	Remaining charge on battery
Individual evidence	Fingerprints
	Drone tail number
Class evidence	Drone material
	DoD class
	Homebuilt or bought
Computer generated evidence	Memory dump
	Storage analysis

The form above covers the following items:

Laptop, cellphone, Tablet, iPad, drone controller – This is an obvious one. If we are able to get ahold of this device, then we could determine ownership.

Maps of crime scene, mapping software, photographs/video/diagrams of crime scene – finding these could indicate the intent. If anything, it could show the crime that was committed. Frequently, criminals post videos of the crimes they committed on YouTube or other social media. You might get lucky and spot a person in the video, which may lead to a name or two. It's also possible to find video on the drone itself. This could even be older video that may lead to who owns the drone

Drone manual, assembly instructions, user manual, pics/videos of drones, custom code – none of this could be considered evidence in itself. However, if you have a drone that was used in a crime and the user manual and/or code that was used to control it, you might have enough evidence to determine who owns it.

Journal entries, threatening evidence, confessions, computer links, and eyewitness testimony – this evidence could tie a person to a drone that might tie them to a crime. Having a taped confession stating exactly what the person's intentions were would prove helpful to determine motive.

Fingerprints, serial numbers, drone manufacturer, make/model – having the drone manufacturer/serial number/make and model, an investigator could ask a drone manufacturer to get drone sales for a specific area and possibly tie it back to a person. As an example, a drone crashed into a crowd at a public park and injured someone. The police now have the drone. It's a DJI Phantom 4 Pro. An investigator could contact DJI and ask if they sold any drones to anyone in the area. Now, there are number of factors here to consider. The drone could be borrowed, the pilot could be from out of town, or the drone could have been resold several times. This is just one avenue the investigator has.

Temperature, weather, date/time, direction the drone was facing when found, and remaining battery charge – none of these can determine drone ownership or the underlying mission of the drone. However, some of these might be indicators. As an example, if a drone was found with 98% battery life, this could indicate that it didn't fly very far and the owner could be nearby.

Weather could indicate why it crashed. It's important to record the weather conditions when arriving on a crash scene. If it is windy or rainy, it could be a factor in why a drone crashed. The direction the drone was found it along with the battery life could help find the owner. Assuming it was found in a northerly facing direction with 98% battery life, one could easily assume it came from a southern direction, but obviously not always.

Even the date and time could tell us something about a drone's owner or mission. What if someone crashed a large drone into a crowd of people on September 11? Is this someone trying to avenge something? What if a drone was involved in an incident at 3:00 am? More than likely an adult is involved and not someone under age, but we are just making blind assumptions here. We would need the full picture in order to make informed decisions.

Fingerprints, drone tail number – Drone tail numbers are easy to look up. There is an Federal Aviation Administration (FAA) database that will show the town and state the drone is registered in. You will have to be in law enforcement in order to find out who actually owns the drone. This information is not released to the general public.

A drone can be dusted for fingerprints. It is doubtful that any matches will come back, but it does put the fingerprints in a documented database. If an arrest is made, then there will be something to compare them against.

Drone material, DoD class, homebuilt versus bought – any of these could help to determine drone ownership. IF a drone was 3D printed for example, asking around the neighborhood if anyone owns a 3D printer might be a good clue to follow. DoD class would indicate if the drone is a recreational or commercial drone. How many individuals own a commercial drone?

Memory analysis – A memory dump of the drones memory will yield quite a bit of information about the drone and possible its owner and true mission. First, dump the memory with DumpIT an open source software application. Memory can then be analyzed with Volatility by the Volatility Foundation, an open source software package that runs on Windows, Linux, or MAC operating systems. During analysis, look for commands that the drone was given. Initial GPS coordinates might still be available along with graphics and flight information.

Storage analysis – Most drones come with some way to store images. Cheaper drones may only hold them in onboard memory and you might need a cable to connect to it to download the storage. We would mainly be looking at images. Images might lead to who owns the drone or where it took off from and what it was photographing, giving way to what its real mission was. Storage analysis can be conducted with a number of tools. Autopsy is one of the easiest to use. This software will recover any deleted data as well. Although not open source, Autopsy is free to download.

Remember that nothing is really deleted in a computer, so doing a forensic analysis could show images that were tagged for deletion and again, lead to who owns the drone. One thing to take note here is that if the drone was resold, you might pick up images from the previous owner. This again could be helpful as they might tell you who they sold it to.

To recover data in Autopsy, use the following directions:

1. Click on Autopsy to start the application
   a. Select New Case
2. A window asking for Case Information will be displayed
   a. Fill in Case Name and Base Directory
   b. click next
3. An Optional window will be displayed
   a. You do not have to fill in anything in this form, but if you are doing a real case, you will want to
   b. A new case will be created
4. The Add Data Source window will be displayed
   a. Choose your data source. If you are analyzing the SD card from a drone, then choose Local Disk. If you took an image, then choose Disk Image or VM File
   b. Click next
5. Choose the disk you want to analyze from the Select Disk button
   a. The rest of the fields are optional
   b. Click next

6. The Configure Ingest Modules will be displayed
    a. Use the default modules chose
    b. Click next
    c. Your image will be ingested into Autopsy
        i. Depending on the size of your disk or image, it could take quite a while to load in the data

LET'S ANALYZE THE DATA

Now that we have collected all of this data, what does it mean? If you work as an investigator, then this section will be pretty easy for you. If you do not, then analysis is hard to teach. Its part art, part science and part "gut" feeling.

As we alluded to in the descriptions above, you are looking for correlations. The simple way to explain this is with an example. You get called to investigate a drone incident. You show up at the scene and are told that a drone was flying in a neighborhood, flying close to houses and looking in windows. One of the officers on scene, found the drone in a field. You connect the drone to your laptop and discover that the drone is owned by someone named "Johnny". You got this information from the "Drone Name: Johnny's drone". An astute investigator would start knocking on doors asking if a Johnny lives in the neighborhood.

This is a very simple example and not all incidents will end this cleanly. Using a more complex example, you get called to investigate a drone flying over a local airport. This concerns the airport owners given the damage a drone can do to an airplane and put people's lives at risk. You arrive at the scene to find the local police have found the drone crashed off the side of a runway. You plug the drone into your laptop and search for a name, but cannot find anything. You notice that some pictures have been taken and download them. You ask the local police of they can pinpoint the area of any of the pictures taken. One policeman says yes, he knows the neighborhood of one of the pictures. He indicates that it's only a few blocks away.

You check the battery level and see that it is only 5% used. This tells you that the pilot must not have been far away. You are taken to the neighborhood the policeman identified and start knocking on doors. It only took three tries until a neighbor identified someone a few doors down has a drone. The scenarios can go on and on and we cannot cover every one. The point here is that you will need to do some work to find out the owner of the drone and what their intention was.

SUMMARY

Due to the increase of people flying drones, there has been an increase in drone incidents; thus, forensics has become increasingly important. This chapter concentrated on drone forensics and what it takes to perform a drone forensic investigation. We have included some of the questions that would need to be asked during the investigation. We have also included some examples of the evidence that would need to be collected by investigators as part of the investigation.

By now you probably have a huge interest in drones. Chapter 11 shows you where to buy a drone, what drone conferences you can attend and where to buy sensor and accessories. Privacy is become a big problem so we have included some states that have passed privacy laws that you should be aware of. We also included a closing note on drones which discusses flying at night, if you add anything to your drone; weight is always a problem, and battery length. Finally, the chapter closes with the future of drones. This is our attempt to predict the evolution of drones five, ten, and even twenty years or more into the future.

Chapter 11

More on drones

Image used with permission from Stephanie. DeFrangesco
DeFrangesco, R. (2021). Quadcopter. Williamstown, NJ
Quadcopter

Now that you are part of the drone community, you will want to grow with this community. This chapter will give you resources in order to enhance your drone's capabilities, attend drone conferences, get a 107 pilots license, and understand the future of drone flight.

It's very important to keep up on drone laws and what other people are doing with drones. the best way to do this is to attend a drone conference. The following are some drone or drone related conferences:

DOI: 10.1201/9781003201533-11

DRONE CONFERENCES

- USDroneCon – This conference used to be held in Maryland, but will be moving to New Jersey. This conference caters mostly to consumers. Check the website for current information at www.usdronecon.com.
- Drone World Expo – This conference is for the commercial user. The conference changes places every year, so check the website for current information at www.expouav.com. It is also known as Commercial Unmanned Aerial Vehicle (UAV) EXPO Americas. This conference is also held in Europe and is called Commercial UAV Expo Europe.
- InterDrone – This conference moves from place to place. You can get updated inforamton from interdrone.com. This conference focuses mostly on commercial uses.
- Xponential – This conference is a conference that focuses on the commercial user. Go to the website for up to date information at www.xponential.org.
- Consumer Electronic Show (CES) – this conference is one of the largest electronic shows in the country. Every aspect of consumer electronics is exhibited from automotive to home uses, to family, and sports. You can get the latest information from www.ces.tech.

Throughout this book, we mention resources for purchasing motors, propellers, transmitters, and drones. This section builds on what was already presented by giving you more places to purchase from.

- Drone manufacturers
  - DJI
  - Parrot Drones SAS
  - Yuneec
  - UVify
  - Hubsan
  - Autel Robotics
  - FreeFly
  - Air Hogs
  - Ambarella
  - 3D Robotics
- Sensor manufacturers
  - LiDAR
    - LeddarTech
    - Velodyne

  - Riegl
  - Routescene

- Ultrasonic sensors

  - MaxBotix
  - FT Technologies
  - Grove Ultrasonic

- Camera manufacturers

  - 4k cameras

    - DJI
    - GoPro
    - Flir
    - Raptor Photonics
    - Teledyne Lumenera
    - Phase One
    - Kappa Optronics

  - Active gimbals

    - DJI
    - Walkera Technology
    - Mio
    - Pixy
    - Gremsy

- Accessories

  - Landing lights

    - LumeCube
    - FlyPro
    - AeroLEDs
    - Aveo Engineering

  - Floatation gear

    - Dronerafts
    - Thekkiinngg
    - Mavic
    - Haoun
    - O'woda

- Batteries
    - Grepow
    - Tattu
    - Lipo
    - Vivitar
    - Fytoo
    - Hoovo
    - Zeee
- Propellers
    - Sensenich
    - Blomiky
    - APC
    - T-Motor
    - RayCorp
    - GIDY
    - Genfan
    - HQProp
    - iFlight
- Backpacks
    - Manfrotto
    - Lykus
    - SSE
    - Seasky
    - Eirmai
    - Mososi
- Toolkits
    - ToolBay
    - Strebito
    - Xool
    - Hobby-Ace
- 3D printer manufacturers
    - Anycubic
    - FDM
    - BIQU
    - Voxelab
    - Creality

- FLSun
- Labists
- Elegoo
- DaVinci

- 3D filament manufacturers

  - Gizmo Dorks
  - NinjaTek
  - Duramic
  - Overture
  - Hatchbox
  - Jarees
  - 3D Solutech

A CLOSING NOTE ON DRONES

Do the right thing as a pilot, fly responsibly. It only takes a few minutes to ruin it for everyone.

Register your drone

Unless you have a very small drone, it's going to need to be registered. Rather than walk the line on what should or should not be registered, just register your drone regardless. It's cheap and relatively painless to register your drone. This shows that you are a drone pilot that cares about the hobby and would like to be part of the drone community.

Weather

If you fly your drone in bad weather including windy, rainy, or snowy conditions and crash into a person or building, you could be liable. Although it's not illegal to fly in these conditions, it's a bad practice to do so. If you cause damages or hurt someone, more than likely, your insurance will not cover you in these circumstances.

Flying at night

It is illegal in most countries to fly your drone at night. Even though you may be able to see your drone's lights at night, you will lose depth perception. If you fly at night and crash your drone, it is highly unlikely that your insurance will cover you. You are breaking the law when you do this and insurance companies generally frown on this type of behavior.

Weight they can lift

Recreational drones are limited in what they can carry. Piling on accessories could cause you to lose maneuverability of your drone and cause it to crash. You are the pilot and you are responsible for making sure your drone is safe to fly. Be careful that if you add accessories to your drone that it can still fly safely and you have total control of it.

Battery length

Drone technology has come a long way. When drones first appeared for recreational use, battery capacity was limited to a few minutes of flying time. Today, some batteries will last almost an hour. The DJI Matrice 300 RTK is a good example of how long a drone can stay flying. It boasts a 55 minute flying time. However, this is the exception and not the rule. Most drones are still limited to 10–20 minutes. The rest are somewhere in between.

Banned countries

We discussed several countries that have banned drone use. This is no joke. If you get caught sneaking your drone into the country and flying it, you run the risk of being arrested and jailed. Not know is not an excuse. The US Embassy may not be able to help you out depending on the country. Know before you fly!

Privacy

We could probably write a book just on privacy. Privacy is a big topic of discussion in the drone community today. It's very easy for a drone pilot, on purpose or by accident, to fly over a neighbor's property and take pictures or video. This infringes on a property owners privacy. Depending on the state you are leaving yourself open for arrest and confiscation of your drone. Respect other people's privacy and don't fly over their property. If your neighbor is having a party, don't fly your drone above their fence line, you are just asking for problems and will ruin it for the rest of the people that fly their drone responsibly.

In 1946, in the United States v. Causby, the courts ruled that, "The landowner owns at least as much of the space above the ground as he can occupy or use in connection with the land". They did not designate how much above the land that is. This has, and remains to be, a subject that is open to interpretation. Legal experts feel that the rights the owner has on their land, extend to the airspace above the land however ill-defined that is.

As an example is the State of Louisiana v. Benson where a drone pilot was arrested and is being charged with intent to surveil. This is a very complicated case in that the pilot followed all of the Federal Aviation Administration (FAA) regulations during flight operations, but he did fly over his neighbor's property. The property owner was a deputy sheriff for the county. The property owner confronted the neighbor, the drone operator, and complained. Less than a week later, the sheriff's department arrested the drone pilot and was charged with an intent to surveil.

Some of the sticky points to this case include the fact that the drone pilot was flying 40 mph at the time over the neighbor's property. This hardly shows an intent to surveil. Video collected from the drone substantiates this fact along with another fact that the pilot did not stop at all over the property. Complicating this case even more, the state of Louisiana is claiming that they have control of the airspace above a property which directly conflicts with congress that states the FAA has control of all airspace. This case has not been resolved yet, but it will be interesting to see how the courts rule.

The following is a list of states and laws pertaining to drones used for surveillance or invasion of privacy:

Arkansas
Act 293: Prohibits the use of drones to commit video voyeurism (invasion of privacy). Class B misdeLmeanor; Class A misdemeanor if images were distributed or transmitted to another party, or posted to the Internet.

Act 1019: Prohibits the use of drones for surveillance and/or the gathering of information on "critical infrastructure" (oil refinery, chemical manufacturing facility, power plants, etc.) without written consent.

California
Civil Code Section 1708.8: Prohibits the use of drones to capture video and/or a sound recording of another person without their consent (invasion of privacy). Violators are liable for up to three times the amount of damages related to the violation, and a civil fine of between $5000 and $50,000.

Florida
Criminal Code Section 934.50: Drones may not be used for surveillance in violation of another party's reasonable expectation of privacy; this includes laws enforcement. However, police may use drones with a valid search warrant. Violators may be ordered to pay legal fees and compensatory damages; victims may seek injunctive relief.

Michigan
Michigan Compiled Laws Section 324.40112: Prohibits the use of drones to harass or interfere with a hunter (charged as a misdemeanor; up to 93 days incarceration and/or up to $1000 fine per offense).

Mississippi
Mississippi Code Section 97-29-61: Prohibits the use of drones to peep into a building for the "lewd, licentious and indecent purpose" of spying on another party (charged as a felony, up to 5 years in prison; up to 10 years in prison if the person spied on is a child 16 or younger)

North Dakota
North Dakota Code Section 29-29.4-01: Limits the use of drones for surveillance, crime investigation, and other uses by law enforcement (creates the requirement of a warrant, etc.)

Tennessee
Tennessee Code Section 39-13-903(a): Prohibits the use of drones to capture images at open-air events where 100 or more people are gathering for a ticketed event (statute specifically refers to fireworks events). The law also prohibits the use of drones over prison grounds.

Texas
Gov. Code Section 423.002(a): Clarifies the legality of using drones to capture images by certain professionals (such as photographers), with the requirement that individuals are not identifiable in images unless they have given express permission.

Utah
Utah Code Title 63G, Chapter 18: Authorizes police to use drones for data collection at testing sites and to find missing persons in areas where there is no reasonable expectation of privacy.

Virginia
Virginia Code Section 19.2-60.1: Requires that police obtain a warrant prior to using a drone for criminal investigations or surveillance (unless it is for an Amber Alert, Senior Alert, or Blue Alert).

Just because your state does not have a specific law regarding privacy, it does not mean that you can't be charged with breaking a law. For instance, let's say you fly your drone over to your neighbor's house to see what he is up to. You fly around the house to a window and accidently see his wife getting changed. Accident or not, this could be considered voyeurism and is prosecutable in most states. The drone has a camera and that's all that is needed in order to charge you.

As an example, a couple from Utah was each charged with one count of voyeurism when they used a drone to film a family in their bedroom and bathroom. One of the victims saw the drone flying outside of his window and decided to get into his car and follow the drone. He tracked the drone to a nearby parking lot and was able to capture the drone and look at the images on the drone's SD card. He saw quite a few images of his family and other people in their bedrooms and bathrooms as well. He turned it into the authorities and they were able to identify the owner.

Utah does not have a specific láw that addresses using a drone for voyeurism. However, in the laws eyes, there is no difference using a drone which does have a camera and a person standing outside the window using a camera to take pictures.

THE FUTURE OF DRONES

The laws governing what drone pilots can and cannot do with their drones will get tighter. The fines will get stiffer and the prison time for breaking the laws will get harsher.

I see drones as going the way of the Segway. When the Segway was released, it was a sensation; everyone wanted one. It was marketed in 2001 as a transportation alternative. Unfortunately, the company ended production of the Segway in 2020. Today, drones are a curiosity to most people. To the average consumer, a drone has little value except for entertainment.

Drone usage will grow in the commercial and military space. Drones will be used as much as possible to replace people doing dangerous or costly jobs.

I expect that the controllers a pilot uses to maneuver a drone will evolve. I can see them voice controlled, much like an Alexa, gesture controlled, or even mind controlled. Using voice and gesture control is in our current capability. I don't see mind control capability for quite a few years yet, but it will come.

Drones will evolve into appliances. Meaning, they will be sold out of the box with certain functions. As an example, if you want to use your drone for entertainment, they will come preprogrammed to do tricks; no clicking of buttons on the controller to do a flip. If you are a commercial company looking to deliver packages, you will be able to purchase drones that do that specific function. The drone will have the claws to hold the package and software that you can easily program an address into.

Connectivity is a real sticking point for drones. Today, you are limited to how far you can fly your drone with limited range and connectivity to the receiving device. On an excellent day, Wi-Fi can only reach 300 feet. This is with the assumption of good atmospheric conditions and no obstructions in the way. Tomorrow's drones will use satellite communications. You will have the capability to control your drone from miles away.

If you remember, when Global Position System (GPS) became available to the general public, it had limited accuracy. Only the government had this capability. The government implemented selective availability, and accuracy was only to within 16 feet. Eventually, the government lifted selected availability, and accuracy improved. Today, you can get a GPS with L5 capability that is accurate to within a foot. Commercial drones will use satellite communication technology to fly miles from the controller. At first, the communication bands will be limited, but the government will likely open the use of satellite communications for drone use.

Unfortunately, consumer drones will not have this capability. Additionally, you will need a whole new class of license to fly using satellite communications. Pilots use to having eyes on their drones, will need to learn to fly using just sensors and cameras. Today's children and young adults will be well suited for this type of work given they are excellent at gaming.

Encryption will become increasingly important to drones in the future. The link must be encrypted so it cannot be intercepted and the drone taken down. Military drones tend to use this feature more, but we have seen the United States lose a drone to a poorly encrypted communications link. Commercial drones will use this in the future. Commercial drones will be bigger and more costly. No one wants to have an expensive drone hijacked or stolen due to an unencrypted communications link.

Battery life has been a real problem with drones from day one. Battery life will be extended into hours in the future. Additionally, you will be able to recharge literally on the fly. You will be able to land your drone on a charging station and within minutes, recharge your battery.

In general, drones will become bigger and capable of more lifting power. This will be needed in order to use a drone as a workhorse. Drones will need to be bigger in order to lift larger cameras along with sensors in order to make them useful.

Commercial drone traffic will increase. Air space will have to be developed to allow for many more commercial drones. Drone ports will be built in order to takeoff and land drones. These will be similar to airports, manned with drone-traffic controllers to direct the takeoff, flight and landing of drones. This also means that radar will have to get better in order to track drones.

Although only a pipe dream today, drones will be used to transport people. This will obviously be a commercial endeavor, but just like planes and helicopters, there will be Do-It-Yourself (DIY) enthusiast that will design and build their own. The FAA will have to allow for airspace for this new class of Unmanned Aerial Systems (UAS's). In order to fly without a pilot, they will have to take advantage of the latest, GPS and obstacle avoidance systems, much like driverless cars do today.

I believe that commercial companies will start out transporting one-to-two people at first, then eventually the passenger numbers will grow to be like a UAS bus. There will be commercial routes they will fly with stops like a city bus. Passengers will get off and on as they wish. This type of UAS will need to be powered by hydrogen cells with backup systems in case of failure.

Finally, we will see huge gains in the military using UAS's as vehicles for transporting people, supplies, and as weaponized platforms. It will start with using them as supply ships because there is still a lot to learn and little to no risk to human life if they crash. They will be armor plated and capable of landing in exact places that require resupply.

The military will weaponize these UAS platforms. It's feasible to mount laser guided weapons to pinpoint where to drop bombs or fire high caliber automatic weapons. A spotter could be used to light up a target and the UAS will fire on that area. Should the UAS be shot down, then it could self-destruct. The key here is that there would be no loss of life. These platforms will be produced by the thousands, so the cost will be minimal to manufacture. This is not a new concept to the military. They are using drones heavily in everything from reconnaissance to firing missiles already, so the jump to using a larger Unmanned Aerial System (UAS) (quadcopter) platform would be minimal.

Don't wait, get out there and start flying!

Definitions

AGL: Above Ground Level. The height that is measured from the ground surface.

Arduino: An open source hardware microcontroller and software platform that allows users to connect to and control the physical world. The Arduino hardware is distributed by the Lesser General Purpose Lucense (LGPL) and General Purpose License (GLP) license, which allows distribution by anyone.

CCW: Counter clockwise (The direction in which one set of propellers spin on a drone).

Code: This is the computer code that we typically call an application.

CW: Clockwise (The direction in which one set of propellers spin on a drone).

DIY: Do-It-Yourself. The act of building or modifying things yourself rather than hiring someone to do them for you.

DoD: Department of Defense. The federal department in charge of national security.

Drone: The FAA defines a drone as a vehicle which is not piloted by a human from within the vehicle itself.

EU: European Union. The EU is made up of 27 countries that are located mostly in Europe. The EU represents almost half a billion people

FAA: Federal Aviation Administration. The FAA is the main authority in the United States that enforces drone laws.

Fixed wing drone: A fixed wing drone is different than a quadcopter in that it typically only has two motors versus four, does not take off vertically, it usually has to be thrown into the air. Most fixed wing drones are designed to look like fighter aircraft.

Forensics: The science that is used to solve a crime. In the drone field, we are looking for who the owner is, why the drone failed, and what was the intent of the pilot.

FPV: First Person View. FPV allows a drone pilot to view video and control a drone from a driver's perspective.

FTP: File Transfer Protocol. A protocol that allows one computer to transfer data to another computer. It should be noted that this protocol is very insecure since it transfers credentials and data in clear text. Secure FTP has replaced this protocol which encrypts the credentials and data.

GHz: Gigahertz is a unit of measure for Alternating Current (AC). It represents one-billion hertz or cycle of time.

GDPR: General Data Protection Regulation. This is the European Union framework for the collection and processing of personal information.

Hacking: The act of trying to gain unauthorized access to a system or computer. In the context of this book, a hacker is someone that modifies their drone to do something it was not designed to do. It could also be someone that uses undocumented features.

IP address: Internet Protocol address. A unique identifier that each computer uses to communicate on a network. IP addresses are also known as logical addresses. There are currently two types of IP addresses; IPv4 and IPv6. IPv4 addresses are 32-bit numbers and typically look like 192.168.1.1. IPv6 addresses are 128-bit and look like 2001:0DB8:AC10:FE01: (the extra zeroes are omitted).

LAANC: Low Altitude Authorization and Notification Capability. A joint venture between industry and the FAA. LAANC provides drone pilots access to controlled airspace below 400 feet.

LiDAR: Light Detection and Ranging. A sensor that works much like radar but uses light from a laser.

MAC address: Media Access Control address. A unique identifier that is assigned by a device manufacturer. Also known as the physical address.

MAh: MilliAmp Hour. One thousandth of an ampere flowing for one hour.

MHz: Megahertz is a unit of measure for Alternating Current (AC). It represents one-million hertz or cycle of time.

NATO: North American Treaty Organization. NATO is really an alliance between roughly 30 countries to provide a collective defense against anyone outside the alliance.

NTSB: National Transportation Safety Board. According to their website, the NTSB is charged with investigating transportation accidents. More than likely, this would be the agency to investigate a drone incident especially if it involved a plane, train, or caused a major accident.

OPSEC: Operational Security. This is the security you apply to protect your equipment, computers, your home and even yourself.

Quadcopter: Also known as a UAV, UAS, or drone. A quadcopter has four motors with propellers that spin to lift the drone.

Raspberry PI: A single board computer originally developed to help teach computer science in schools. The PI was developed by the Raspberry Pi Foundation along with Broadcom. Current models include the Pi Zero (W and WH), Pi 4 Model B, Pi 400, and Pi Pico. The boards are popular with hobbyist, starting as little as $5 USD for the Pi Zero.

Sensor: In the context of drone technology, a sensor is any device that attaches to the drone and collects physical data to be processed. As an example, this could be lidar, radar, or a thermal imaging camera.

Telnet: A protocol used to connect to a remote computer. This is an outdated protocol. SSH (Secure Shell Protocol) should be used when feasible. Most drones support Telnet, but not all support SSH.

TSA: Transportation Safety Administration. Created after 9/11, the TSA is tasked with strengthening the security of our nation's transportation systems.

UAS: Unmanned Aerial System. Another name for a UAV, quad, or a drone.

UAV: Unmanned Aerial Vehicle. Another name for a UAS, quad, or a drone.

Index

B

Building your own drone
  Build your own drone, 115
  DIY drone kits, 116
  3D printing, 117
  Drone racing, 115
  Underwater drone, 121

C

A closing note on drones, 155

D

Definitions, 163
DIY Drone Projects
  Accessories, 126
  DIY Drone projects, 125
  Drone bug-out bag, 132
  A drone light show, 136
  Drone race course, 130
  Drone *versus* drone, 134
  How to make money with a drone, 137
  Multiple drones, 133
  Pwnagotchi, 135
Drone Forensics
  Analyze the data, 148
  Drone forensics, 141
  Introduction to forensics, 142
  Key questions, 143

F

Flying a drone
  Auto takeoff/land, 75
  Commercial flying, 72
  Drone incidents, 77
  Drone maintenance, 70
  Flying a drone, 65
  FPV, 73
  How to build an obstacle course, 76
  Know your drone, 68
  Military drone pilots, 80
  Obstacle Avoidance, 74
  Registration, 67
  Return To Home, 74
  Safety tips, 66
  Spare Parts, 70
  Stunts, 75
  UA visual perception, 78
  Using headless mode, 73

H

Hacking a drone
  Drone Identification, 89
  Hacking a drone, 81
  MAC address, 85
  Protecting your drone, 99
  Replay attack, 88
  Taking down a drone, 97
  Telnet, 82
  Transferring files, 84
Hardware and Software
  Commercial Drone Manufacturers, 50
  Drone add-ons and accessories, 56
  Drone Costs, 55
  Hardware and Software, 45
  Military Drones, 53
  Troubleshooting, 59
History of Drones
  Ancient times, 16
  Chapter Two Summary, 27
  Fixed wing, 23
  History of Drones, 15

The history of the FAA, 26
Iran, 22
Military drones use, 22
Other countries own drones, 23
Remote controlled, 23
Takeoff and landing, 23
Use of military drones today, 22
WWI, 18
WWII, 19

I

Introduction
Agriculture, 5
Chapter One Summary, 13
Cinematography, 5
Construction, 5
DoD drone classes, 9
Drone delivery, 6
The Drone Market, 8
Drone racing, 7
Entertainment, 6
Finding people, 6
Fun, 7
Introduction, 1
Military, 7
Other classification systems, 10
Save lives, 6

L

Laws Governing Drones
Airports, 30
Drone accidents, 37
International laws, 40
Laws governing drones, 29
National Parks, 33
Part 107, 35
Privacy, 39
Register your drone, 36
Sport Arenas, 31
The White House, 30
Wildlife refuges, 31

M

More on drones
Accessories, 153
Camera manufacturers, 153
Drone conferences, 152
Drone manufacturers, 152
The future of drones, 159
More on drones, 151
Sensor manufacturers, 152
States and laws pertaining to drones used for surveillance or invasion of privacy, 157

P

Programming a drone
Code Explanation, 105
Programming a drone, 103
Python, 108

Printed in the United States
by Baker & Taylor Publisher Services

Printed in the United States
by Baker & Taylor Publisher Services